CAMBRIDGE TRACTS IN MATHEMATICS

General Editors

B. BOLLOBÁS, W. FULTON, A. KATOK, F. KIRWAN,
P. SARNAK, B. SIMON, B. TOTARO

178 Analysis in Positive Characteristic

Analysis in Positive Characteristic

ANATOLY N. KOCHUBEI

National Academy of Sciences of Ukraine

CAMBRIDGE UNIVERSITY PRESS
Cambridge, New York, Melbourne, Madrid, Cape Town, Singapore, São Paulo, Delhi

Cambridge University Press
The Edinburgh Building, Cambridge CB2 8RU, UK

Published in the United States of America by Cambridge University Press, New York

www.cambridge.org
Information on this title: www.cambridge.org/9780521509770

© A. Kochubei, 2009

This publication is in copyright. Subject to statutory exception
and to the provisions of relevant collective licensing agreements,
no reproduction of any part may take place without
the written permission of Cambridge University Press.

First published 2009

Printed in the United Kingdom at the University Press, Cambridge

A catalogue record for this publication is available from the British Library

ISBN 978-0-521-50977-0 hardback

Cambridge University Press has no responsibility for the persistence or
accuracy of URLs for external or third-party internet websites referred to
in this publication, and does not guarantee that any content on such
websites is, or will remain, accurate or appropriate.

Contents

Preface		vii
1	**Orthonormal systems and their applications**	1
	1.1 Basic notions	1
	1.2 Additive Carlitz polynomials	13
	1.3 Hyperdifferentiations	17
	1.4 The digit principle	22
	1.5 Finite places of a global function field	32
	1.6 The Carlitz module	36
	1.7 Canonical commutation relations	40
	1.8 Comments	44
2	**Calculus**	47
	2.1 $\mathbb{F}_q$-linear calculus	47
	2.2 Umbral calculus	60
	2.3 Locally analytic functions	75
	2.4 General smooth functions	82
	2.5 Entire functions	90
	2.6 Measures and divided power series	99
	2.7 Comments	102
3	**Differential equations**	105
	3.1 Existence and uniqueness theorems	105
	3.2 Strongly nonlinear equations	114
	3.3 Regular singularity	120
	3.4 Evolution equations	131
	3.5 Comments	135
4	**Special functions**	137
	4.1 Hypergeometric functions	137

	4.2	Analogs of the Bessel functions and Jacobi polynomials	142
	4.3	Polylogarithms	145
	4.4	K-binomial coefficients	154
	4.5	Overconvergence properties	159
	4.6	Comments	166
5	**The Carlitz rings**		167
	5.1	Algebraic preliminaries	167
	5.2	The Carlitz rings	173
	5.3	The ring $\mathfrak{A}_1$	178
	5.4	Quasi-holonomic modules	185
	5.5	Comments	201
Bibliography			203
Index			209

Preface

One of the natural options in the development of mathematical analysis is to investigate analogs of classical objects in a new environment. To obtain the latter, it suffices to change the ground field. For a nontrivial analysis, we need a nondiscrete topological field. If we confine ourselves to locally compact fields (which can be extended subsequently via algebraic closures and completions), then the spectrum of possibilities is not very wide – $\mathbb{R}$, $\mathbb{C}$, the fields $\mathbb{Q}_p$ of p-adic numbers and their finite extensions, and non-Archimedean local fields of positive characteristic, that is the fields of formal Laurent series with coefficients from finite fields $\mathbb{F}_q$.

Over $\mathbb{R}$ and $\mathbb{C}$, the whole of classical mathematics is developed. The p-adic fields constitute the environment for p-adic analysis, a rapidly developing branch of contemporary mathematics embracing analogs of classical function theory, Fourier analysis, differential equations, theory of dynamical systems, probability, etc. Note that there are two kinds of p-adic analysis. One of them deals essentially with functions from $\mathbb{Q}_p$ to $\mathbb{R}$ or $\mathbb{C}$, like characters or probability densities, and its main results remain valid for local fields of positive characteristic (see [59]). The second treats functions with non-Archimedean arguments and values, like polynomials and power series, and here the extension to fields of positive characteristic is much more complicated.

To see the essence of these difficulties, it suffices to notice that some of the simplest standard notions of analysis do not make sense in the case of characteristic $p > 0$. For example, $n! = 0$ in a field of characteristic p, if $n \geq p$. Similarly, $\frac{d}{dt}(t^n) = 0$ if p divides n, that is the classical differential calculus cannot be used to investigate important classes of functions. For the same reason, the standard formulas do not work if one wants to define analogs of the most important special functions, beginning with the exponential function.

The first steps in developing analysis over local fields of positive characteristic were made in 1935 by Carlitz [22] who found appropriate positive characteristic counterparts for the factorial, exponential, logarithm, and classical orthogonal polynomials. Carlitz's constructions were highly unusual for that time, and their algebraic meaning was understood only about 40 years later, within the theory of Drinfeld modules and their generalizations.

After that they became important elements of a branch of number theory, function field arithmetic; see the books by Goss [45], Rosen [94], and Thakur [111].

Though the initial paper by Carlitz contained important insights into analysis (like the difference operator, very close to a later notion of the Carlitz derivative), the active development of analysis in positive characteristic began much later, after the paper [43] by Goss (with an exposition of Carlitz's work in more modern language) and the paper [109] by Thakur who introduced analogs of the hypergeometric function and the hypergeometric differential equation. That opened the way for a systematic development of the counterparts of the theory of Fourier series, the theory of differential equations, etc. The main achievements in these and related directions are summarized in this book.

In contrast to the books by Goss [45] and Thakur [111], whose material has some common features with the present one, this book is intended for specialists in analysis, from the graduate student level. Therefore only the most basic knowledge of algebra is presumed; in any case, an explanation of any algebraic notion, not given explicitly in the text, can be found in Lang's *Algebra*. In order to make the material more elementary, most notions are introduced on the local field level (though in some cases their global field interpretation is also outlined). The author's aim was to reach analysts who do not usually read books or papers in algebraic number theory and algebraic geometry. Accordingly, the author did not try to touch on some nonelementary material covered exhaustively in [45, 111], like the zeta and gamma functions and their arithmetic applications (a new analog of the zeta function, appearing in Chapter 4, is still only an example of an application of differential equations with the Carlitz derivatives). Some results requiring involved algebraic techniques are given without proofs.

Here is a short description of the contents. In Chapter 1, we introduce and study systems of functions on a local field of positive characteristic, which are crucial for solving various specific problems of analysis – the additive Carlitz polynomials and hyperdifferentiations. These functions are $\mathbb{F}_q$-linear; a method of lifting a $\mathbb{F}_q$-linear basis to a basis of a space of continuous functions was invented by Carlitz (1940) and found to be actually a very general construction ("the digit principle") by Conrad (2000). We describe both the general results and some specific cases. Then we consider the first special functions, the Carlitz exponential and logarithm, and the Carlitz module, and show how the above material appears in the positive characteristic analog of canonical commutation relations of quantum mechanics.

Chapter 2 deals mainly with the representation of functions by their expansions in a series of Carlitz polynomials, the correct understanding of the smoothness and analyticity of functions and their characterizations in terms of Fourier–Carlitz coefficients, the first notions of differential and integral calculus, which agree (in contrast to the classical ones) with the positive characteristic framework. The differential calculus is then extended to a kind of Rota's umbral calculus.

Given the notion of a derivative, it is natural to study differential equations, and that is done in Chapter 3, including analogs of Cauchy's existence and uniqueness theorems for analytic solutions, counterparts of regular singularity theory and partial differential equations of evolution type.

Some special functions satisfying differential equations with Carlitz derivatives are discussed in Chapter 4. Finally, in Chapter 5 we introduce and study rings of differential operators of the above kind. These rings can be seen as analogs of the Weyl algebras, though some of their properties are quite different – in fact, already the Carlitz derivative is, for instance, nonlinear (actually, $\mathbb{F}_q$-linear). Nevertheless, there exists a notion of quasi-holonomic modules over these rings, having some common features with holonomic modules in the sense of Bernstein, and connected to some special functions in the spirit of Zeilberger's theory.

The author is grateful to D. Goss, W. H. Schikhof and D. S. Thakur for their helpful consultations in non-Archimedean analysis and function field arithmetic.

Anatoly N. Kochubei

1
Orthonormal systems and their applications

In this chapter we introduce the main notion of analysis over a local field K of positive characteristic, and study two most important orthonormal systems appearing in this theory, the Carlitz polynomials and hyperdifferentiations. Their role, especially that of the Carlitz polynomials, will be crucial throughout this book. However the first applications, to the Carlitz module expressing the main functional relation for the counterparts of the exponential and logarithm, and to representations over K of the canonical commutation relations of quantum mechanics, will be given in this chapter. We also describe the digit principle connecting the analysis of additive functions and that of general continuous functions on subsets of K.

1.1 Basic notions

1.1.1. Function fields. Factorials. It is well known [17, 38, 122] that any local field (= nondiscrete locally compact topological field) of positive characteristic p is isomorphic to the field K of formal Laurent series with coefficients from the Galois field $\mathbb{F}_q$ of q elements, $q = p^v$, $v \in \mathbb{N}$. There is an absolute value $|\cdot|$ on K defined as follows: $|0| = 0$; if $z \in K$,

$$z = \sum_{i=m}^{\infty} \zeta_i x^i, \quad m \in \mathbb{Z}, \ \zeta_i \in \mathbb{F}_q, \ \zeta_m \neq 0,$$

then $|z| = q^{-m}$. The absolute value has the following properties:

$$|z_1 z_2| = |z_1| \cdot |z_2|, \quad |z_1 + z_2| \leq \max(|z_1|, |z_2|), \quad z_1, z_2 \in K.$$

The second property, called *the ultra-metric inequality* (or *the non-Archimedean property*), implies that K is totally disconnected in the topology defined by the metric $|x - y|$. Here we encounter all the peculiar properties of non-Archimedean spaces (see, for example, [98]) – standard subsets of K, like balls or spheres, are simultaneously open and closed (in fact, compact); two balls either do not intersect, or one of them is contained in the other, a series $\sum_{k=1}^{\infty} a_k$, $a_k \in K$, converges if and only if $a_k \to 0$, etc.

We will denote by $\overline{K}$ a fixed algebraic closure of K. The absolute value $|\cdot|$ can be extended in a unique way onto $\overline{K}$, and the completion $\overline{K}_c$ is algebraically closed (see [98]).

The first "appropriate" analog of the factorial is the sequence

$$D_i = [i][i-1]^q \ldots [1]^{q^{i-1}}, \quad [i] = x^{q^i} - x \ (i \geq 1), \ D_0 = 1, \tag{1.1}$$

Another factorial-like sequence is

$$L_i = [i][i-1] \ldots [1] \ (i \geq 1), \ L_0 = 1. \tag{1.2}$$

It is easy to see that

$$|D_i| = q^{-\frac{q^i-1}{q-1}}, \quad |L_i| = q^{-i}. \tag{1.3}$$

As we will see, in some cases it is natural to consider D_i as a counterpart of the factorial of the number q^i. Then analogs of other values of the factorial are defined as follows. Let us write any natural number j in the base q as

$$j = \sum_{i=0}^{n-1} \alpha_i q^i, \quad \alpha_i \in \mathbb{Z}, \ 0 \leq \alpha_i < q.$$

Then we set

$$\Gamma_j = \prod_{i=0}^{n-1} D_i^{\alpha_i}. \tag{1.4}$$

Lemma 1.1 (i) *The elements D_m, L_m, and Γ_{q^m-1} are connected by the identity*

$$\Gamma_{q^m-1} L_m = D_m.$$

(ii) *For any $i = 1, 2, \ldots$, the element $[i]$ is the product of all monic irreducible polynomials $m \in \mathbb{F}_q[x]$, such that $\deg m$ divides i.*

(iii) For any $i = 1, 2, \ldots$, the element D_i is the product of all monic polynomials $g \in \mathbb{F}_q[x]$ with $\deg g = i$.

(iv) L_i is the least common multiple of all polynomials from $\mathbb{F}_q$ of degree i.

Proof. (i) By the definition,
$$\Gamma_{q^m-1} = (D_0 \ldots D_{m-1})^{q-1}.$$

We have $D_m = \prod_{i=1}^{m} [i]^{q^{m-i}}$, $L_m = \prod_{i=1}^{m} [i]$,

$$D_0 \ldots D_{m-1} = \prod_{j=1}^{m-1} \prod_{i=1}^{j} [i]^{q^{j-i}} = \prod_{i=1}^{m-1} \prod_{j=i}^{m-1} [i]^{q^{j-i}}$$
$$= \prod_{i=1}^{m-1} [i]^{\sum_{j=i}^{m-1} q^{j-i}} = \prod_{i=1}^{m-1} [i]^{\frac{q^{m-i}-1}{q-1}},$$

so that
$$\Gamma_{q^m-1} L_m = [m] \prod_{i=1}^{m-1} [i]^{q^{m-i}} = D_m.$$

(ii) Let $m \in \mathbb{F}_q[x]$ be a monic irreducible polynomial. Then m divides $[i] = x^{q^i} - x$ if and only if $\deg m$ divides i.

Indeed, suppose that m divides $[i]$. Let α be a root of m lying in the splitting field of m. Then $\alpha^{q^i} = \alpha$, so that $\alpha \in \mathbb{F}_{q^i}$. Therefore the simple extension $\mathbb{F}_q(\alpha)$ of the field $\mathbb{F}_q$ is a subfield of $\mathbb{F}_{q^i}$. Since m is irreducible, we have $[\mathbb{F}_q(\alpha) : \mathbb{F}_q] = \deg m$, and it follows from the basic properties of finite fields [75, 76] that $\deg m$ divides $i = [\mathbb{F}_{q^i} : \mathbb{F}_q]$.

Conversely, if the number $l = \deg m$ divides i, then [75, 76] $\mathbb{F}_{q^l}$ is a subfield of $\mathbb{F}_{q^i}$. If α is a root of m in its splitting field over $\mathbb{F}_q$, then $[\mathbb{F}_q(\alpha) : \mathbb{F}_q] = l$, whence $\mathbb{F}_q(\alpha) = \mathbb{F}_{q^l}$. Therefore $\alpha \in \mathbb{F}_{q^i}$, so that $\alpha^{q^i} = \alpha$, and α is a root of the polynomial $x^{q^i} - x$. Since m is irreducible, it generates the principal ideal in $\mathbb{F}_q[x]$ consisting of polynomials for which α is a root. This means that m divides $x^{q^i} - x$.

Now it remains to note that $\frac{d}{dx}[i] = -1$, so that $[i] = x^{q^i} - x$ has no multiple roots in its splitting field over $\mathbb{F}_q$. Therefore each monic irreducible polynomial appears exactly once in the canonical decomposition of $[i]$ in the ring $\mathbb{F}_q[x]$.

(iii) Let $\pi \in \mathbb{F}_q[x]$ be an irreducible monic polynomial, $\deg \pi = d$. Among all monic polynomials $g \in \mathbb{F}_q[x]$ of degree $i \geq d$, q^{i-d} polynomials are multiples of π, q^{i-2d} ($i \geq 2d$) polynomials are multiples of π^2, etc. Thus the number of polynomials of degree i, whose canonical decompositions contain π with exactly the power 1, equals $q^{i-d} - q^{i-2d}$. We can perform similar evaluations with higher degrees of π, so that the product of all monic polynomials of the degree i contains π with the power equal to

$$\left(q^{i-d} - q^{i-2d}\right) + 2\left(q^{i-2d} - q^{i-3d}\right) + \cdots + \mathrm{int}\left(\frac{i}{d}\right)q^{i-\mathrm{int}\left(\frac{i}{d}\right)d} = \sum_{\nu=1}^{\mathrm{int}\left(\frac{i}{d}\right)} q^{i-\nu d}$$

where $\mathrm{int}(s)$, $s \in \mathbb{R}^+$, means the biggest integer not exceeding s.

On the other hand, considering D_i we see that π is contained once in the canonical decomposition of $[j]$ exactly for $j = \nu d$, $\nu \leq \mathrm{int}\left(\frac{i}{d}\right)$. Since $D_i = \prod_{j=1}^{i} [j]^{q^{i-j}}$, we find that π is contained in the canonical decomposition of D_i with the power $\sum_{\nu=1}^{\mathrm{int}\left(\frac{i}{d}\right)} q^{i-\nu d}$, the same as the above one.

(iv) Reasoning similar to (iii) shows that both elements to be proved equal contain π with the same power $\mathrm{int}\left(\frac{i}{d}\right)$. ∎

1.1.2. $\mathbb{F}_q$**-linear functions.** A special feature of the field K is the availability of many functions f with the property of $\mathbb{F}_q$-*linearity*

$$f(t_1 + t_2) = f(t_1) + f(t_2), \quad f(\xi t) = \xi f(t),$$

where t, t_1, t_2 belong to a $\mathbb{F}_q$-subspace of K, $\xi \in \mathbb{F}_q$. Such are, for example, all power series $\sum c_k t^{q^k}$, $c_k \in \overline{K}_c$, convergent on some region in K or $\overline{K}_c$.

In particular, let us consider $\mathbb{F}_q$-linear polynomials. Such a polynomial over $\overline{K}_c$ (and over any infinite field) has the form $\sum c_k t^{q^k}$ [45]. Obviously, the roots of an $\mathbb{F}_q$-linear polynomial form an $\mathbb{F}_q$-vector space. For separable polynomials, the converse is also true.

Proposition 1.2 *A separable polynomial $P \in \overline{K}_c[t]$ is $\mathbb{F}_q$-linear if and only if the set $W = \{w_1, \ldots, w_m\}$ of all its roots forms an $\mathbb{F}_q$-vector subspace of $\overline{K}_c$*

Proof. The necessity is obvious, and we prove the sufficiency. We have

$$P(t) = \prod_{i=1}^{m} (t - w_i).$$

It is clear that $P(t+w) = P(t)$ for any $w \in W$. Suppose that $y \in \overline{K}_c$. Set
$$H(t) = P(t+y) - P(t) - P(y).$$
Then $\deg H < \deg P = m$, and at the same time $H(w) = 0$ for all $w \in W$. This means that $H(t) = 0$ for all t, so that H is additive.

Similarly, for $\alpha \in \mathbb{F}_q$ set
$$H_\alpha(t) = P(\alpha t) - \alpha P(t).$$
Taking a basis in the finite $\mathbb{F}_q$-vector space W we see that $m = q^l$ for some $l \in \mathbb{Z}_+$. Then $\deg P = q^l$. Since $\alpha^{q^l} = \alpha$, we conclude that $\deg H_\alpha < q^l$. On the other hand, $H_\alpha(w) = 0$ for $w \in W$. Therefore $H_\alpha(t) \equiv 0$, as desired. ∎

A detailed exposition of algebraic properties of $\mathbb{F}_q$-linear polynomials is given in [45]. Here we describe some properties of $\mathbb{F}_q$-linear power series convergent on a neighborhood of the origin.

Let $\mathcal{R}_K$ be the set of all formal power series $a = \sum_{k=0}^{\infty} a_k t^{q^k}$ where $a_k \in K$, $|a_k| \leq A^{q^k}$, and A is a positive constant depending on a. In fact each series $a = a(t)$ from $\mathcal{R}_K$ converges on a neighborhood of the origin in K (and $\overline{K}_c$).

$\mathcal{R}_K$ is a ring with respect to the termwise addition and the composition
$$a \circ b = \sum_{l=0}^{\infty} \left(\sum_{n=0}^{l} a_n b_{l-n}^{q^n} \right) t^{q^l}, \quad b = \sum_{k=0}^{\infty} b_k t^{q^k},$$
as the operation of multiplication. Indeed, if $|b_k| \leq B^{q^k}$, then, by the ultra-metric property of the absolute value,
$$\left| \sum_{n=0}^{l} a_n b_{l-n}^{q^n} \right| \leq \max_{0 \leq n \leq l} A^{q^n} \left(B^{q^{l-n}} \right)^{q^n} \leq C^{q^l}$$
where $C = B \max(A, 1)$. The unit element in $\mathcal{R}_K$ is $a(t) = t$. It is easy to check that $\mathcal{R}_K$ has no zero divisors.

If $a \in \mathcal{R}_K$, $a = \sum_{k=0}^{\infty} a_k t^{q^k}$, is such that $|a_0| \leq 1$ and $|a_k| \leq A^{q^k}$, $|A| \geq 1$, for all k, then we may write
$$|a_k| \leq A_1^{q^k - 1}, \quad k = 0, 1, 2, \ldots,$$
if we take $A_1 \geq A^{q^k/(q^k-1)}$ for all $k \geq 1$. If also $b = \sum_{k=0}^{\infty} b_k t^{q^k}$, $|b_k| \leq B_1^{q^k - 1}$,

$B_1 \geq 1$, then for $a \circ b = \sum_{l=0}^{\infty} c_l t^{q^l}$ we have

$$|c_l| \leq \max_{i+j=l} A_1^{q^i-1} \left(B_1^{q^j-1}\right)^{q^i} \leq C_1^{q^l-1}$$

where $C_1 = \max(A_1, B_1)$. In particular, in this case the coefficients of the series for a^n (the composition power) satisfy an estimate of this kind, with a constant independent of n.

Proposition 1.3 *The ring $\mathcal{R}_K$ is a left Ore ring, thus it possesses a classical ring of fractions.*

Proof. By Ore's theorem (see [50]) it suffices to show that for any elements $a, b \in \mathcal{R}_K$ there exist elements $a', b' \in \mathcal{R}_K$ such that $b' \neq 0$ and

$$a' \circ b = b' \circ a. \tag{1.5}$$

We may assume that $a \neq 0$,

$$a = \sum_{k=m}^{\infty} a_k t^{q^k}, \quad b = \sum_{k=l}^{\infty} b_k t^{q^k},$$

$m, l \geq 0$, $a_m \neq 0$, $b_l \neq 0$.

Without restricting generality we may assume that $l = m$ (if we prove (1.5) for this case and if, for example, $l < m$, we set $b_1 = t^{q^{m-l}} \circ b$, find a'', b' in such a way that $a'' \circ b_1 = b' \circ a$, and then set $a' = a'' \circ t^{q^{m-l}}$), and that $a_l = b_l = \alpha$, so that

$$a = \alpha t^{q^l} + \sum_{k=l+1}^{\infty} a_k t^{q^k}, \quad b = \alpha t^{q^l} + \sum_{k=l+1}^{\infty} b_k t^{q^k},$$

$\alpha \neq 0$.

We seek a', b' in the form

$$a' = \sum_{j=l}^{\infty} a'_j t^{q^j}, \quad b' = \sum_{j=l}^{\infty} b'_j t^{q^j}.$$

The coefficients a'_j, b'_j can be defined inductively. Set $a'_l = b'_l = 1$. If a'_j, b'_j have been determined for $l \leq j \leq k-1$, then a'_k, b'_k are determined from the equality of the $(k+l)$-th terms of the composition products:

$$a'_k \alpha^{q^k} + \sum_{\substack{i+j=k+l \\ j \neq l}} a'_i b_j^{q^i} = b'_k \alpha^{q^k} + \sum_{\substack{i+j=k+l \\ j \neq l}} b'_i a_j^{q^i}$$

(the above sums do not contain nontrivial terms with a'_i, b'_i, $i \geq k$, since $a_j = b_j = 0$ for $j < l$).

In particular, we may set $b'_k = 0$,

$$a'_k = \alpha^{-q^k} \left\{ \sum_{\substack{i+j=k+l \\ i<k, j\neq l}} \left(a'_i b_j^{q^i} - b'_i a_j^{q^i} \right) \right\}.$$

If this choice is made for each $k \geq l+1$, then we have $b'_i = 0$ for every $i \geq l+1$, so that

$$a'_k = \alpha^{-q^k} \sum_{\substack{i+j=k+l \\ i<k, j\neq l}} a'_i b_j^{q^i}. \tag{1.6}$$

Denote $C_1 = |\alpha|^{-1}$. We have $|b_j| \leq C_2^{q^j}$ for all j. Denote, further, $C_3 = \max(1, C_1, C_2)$, $C_4 = C_3^{q^{l+2}}$. Let us prove that

$$|a'_k| \leq C_4^{q^k}.$$

Suppose that $|a'_i| \leq C_4^{q^i}$ for all i, $l \leq i \leq k-1$ (this is obvious for $i = l$, since $a'_l = 1$). By (1.6),

$$|a'_k| \leq C_1^{q^k} \max_{\substack{i+j=k+l \\ i<k, j\neq l}} C_4^{q^i} C_2^{q^{i+j}} \leq C_1^{q^k} C_4^{q^{k-1}} C_2^{q^{k+l}}$$

$$\leq C_3^{q^k + q^{k+l+1} + q^{k+l}} = C_3^{(1+q^l+q^{l+1})q^k} \leq \left(C_3^{q^{l+2}} \right)^{q^k} = C_4^{q^k},$$

as desired. Thus $a' \in \mathcal{R}_K$. ∎

Every nonzero element of $\mathcal{R}_K$ is invertible in the ring of fractions $\mathcal{A}_K$, which is actually a skew field consisting of formal fractions $c^{-1}d$, $c, d \in \mathcal{R}_K$.

Proposition 1.4 *Each element $a = c^{-1}d \in \mathcal{A}_K$ can be represented in the form $a = t^{q^{-m}} a'$ where $t^{q^{-m}}$ is the inverse of t^{q^m}, $a' \in \mathcal{R}_K$.*

Proof. It is sufficient to prove that any nonzero element $c \in \mathcal{R}_K$ can be written as $c = c' \circ t^{q^m}$ where c is invertible in $\mathcal{R}_K$.

Let $c = \sum_{k=m}^{\infty} c_k t^{q^k}$, $c_m \neq 0$, $|c_k| \leq C^{q^k}$. Then

$$c = c_m \left(t + \sum_{l=1}^{\infty} c_m^{-1} c_{m+l} t^{q^l} \right) \circ t^{q^m}$$

where $\left|c_m^{-1} c_{m+l}\right| \leq C_1^{q^l-1}$ for all $l \geq 1$, if C_1 is sufficiently large. Denote

$$w = \sum_{l=1}^{\infty} c_m^{-1} c_{m+l} t^{q^l}, \quad c' = c_m(t+w).$$

The series $(t+w)^{-1} = \sum_{n=0}^{\infty} (-1)^n w^n$ converges in the standard non-Archimedean topology of formal power series (see [83], Sect. 19.7) because the formal power series for w^n begins from the term with t^{q^n}; recall that w^n is the composition power, and t is the unit element. Moreover, $w^n = \sum_{j=n}^{\infty} a_j^{(n)} t^{q^j}$ where $\left|a_j^{(n)}\right| \leq C_1^{q^j-1}$ for all j, with the same constant independent of n. Using the ultra-metric inequality we find that the coefficients of the formal power series $(t+w)^{-1} = \sum_{j=0}^{\infty} a_j t^{q^j}$ (each of them is, up to a sign, a finite sum of the coefficients $a_j^{(n)}$) satisfy the same estimate. Therefore $(c')^{-1} \in \mathcal{R}_K$. ∎

The skew field of fractions $\mathcal{A}_K$ can be imbedded into wider skew fields where operations are more explicit. Let K_{perf} be the perfect closure of the field K. Denote by $\mathcal{A}_{K_{\text{perf}}}^{\infty}$ the composition ring of $\mathbb{F}_q$-linear formal Laurent series $a = \sum_{k=m}^{\infty} a_k t^{q^k}$, $m \in \mathbb{Z}$, $a_k \in K_{\text{perf}}$, $a_m \neq 0$ (if $a \neq 0$). Since τ is an automorphism of K_{perf}, $\mathcal{A}_{K_{\text{perf}}}^{\infty}$ is a special case of the well-known ring of twisted Laurent series [83]. Therefore $\mathcal{A}_{K_{\text{perf}}}^{\infty}$ is a skew field.

Let $\mathcal{A}_{K_{\text{perf}}}$ be a subring of $\mathcal{A}_{K_{\text{perf}}}^{\infty}$ consisting of formal series with $|a_k| \leq A^{q^k}$ for all $k \geq 0$. Just as in the proof of Proposition 1.4, we show that $\mathcal{A}_{K_{\text{perf}}}$ is actually a skew field. Its elements can be written in the form $t^{q^{-m}} \circ c$ where c is an invertible element of the ring $\mathcal{R}_{K_{\text{perf}}} \in \mathcal{A}_{K_{\text{perf}}}$ of formal power series $\sum_{k=0}^{\infty} a_k t^{q^k}$. In contrast to the case of the skew field $\mathcal{A}_K$, in $\mathcal{A}_{K_{\text{perf}}}$ the multiplication of $t^{q^{-m}}$ by c is indeed the composition of (locally defined) functions, so that $\mathcal{A}_{K_{\text{perf}}}$ consists of fractional power series understood in the classical sense.

Of course, $\mathcal{A}_{K_{\text{perf}}}$ can be extended further, by considering $\overline{K}$ or $\overline{K}_c$ instead of K_{perf}. The above reasoning carries over to these cases (we can also consider the ring $\mathcal{R}_{\overline{K}_c}$ of locally convergent $\mathbb{F}_q$-linear power series as the initial ring). In each of them the presence of a fractional composition factor $t^{q^{-m}}$ is an $\mathbb{F}_q$-linear counterpart of a pole of order m. Thus we may see the above skew fields as consisting of functions of meromorphic type.

1.1.3. Orthonormal bases. In the sequel we will need some elementary notions from non-Archimedean functional analysis. The reader can consult [93, 100, 82] for further information on this rich and well-developed subject.

Let E be a vector space over K (similarly one can deal with any non-Archimedean local field). A norm on E is a map $u \mapsto \|u\|$ from E to $\mathbb{R}$, such that for $u, v \in E$, $\lambda \in K$,

$$\|u\| \geq 0, \quad \|\lambda u\| = |\lambda| \cdot \|u\|, \quad \|u + v\| \leq \max(\|u\|, \|v\|),$$

and $\|u\| = 0$ if and only if $u = 0$. If E is complete as a metric space with the metric $(u, v) \mapsto \|u - v\|$, then the space E is called *a Banach space* over K.

In this book we will often deal with the Banach space $C(O, K)$ of all K-valued continuous functions on the ring of integers $O = \{z \in K : |z| \leq 1\}$, equipped with the norm

$$\|u\| = \sup_{s \in O} |u(s)|,$$

and its subspace $C_0(O, K)$ consisting of $\mathbb{F}_q$-linear functions.

A sequence $\{f_0, f_1, f_2, \ldots\}$ in a Banach space E over K is called *an orthonormal basis*, if each $u \in E$ can be represented as a convergent series

$$u = \sum_{n=0}^{\infty} c_n f_n$$

where $c_n \in K$, $c_n \to 0$, and

$$\|u\| = \max_{n \geq 0} |c_n|.$$

The coefficients in such a representation are determined in a unique way.

Denote by $P \subset K$ *the prime ideal*

$$P = xO = \{z \in K : |z| < 1\}.$$

Then *the residue field* O/P is isomorphic to $\mathbb{F}_q$. If $E_0 = \{u \in E : \|u\| \leq 1\}$, then the *residual space* (or *the reduction*) $\overline{E} = E_0/PE_0$ is a vector space over the residue field O/P. In an obvious way, any element from E_0 possesses a reduction from $\overline{E}$.

Denote by $\|E\|$ and $|K|$ the sets of values of the norm $\|u\|$, $u \in E$, and of the absolute value $|z|, z \in K$, respectively.

Proposition 1.5 *Suppose that* $\|E\| = |K|$. *A sequence* $\{f_n\} \subset E$ *is an orthonormal basis in* E, *if and only if all the vectors* f_n *lie in* E_0, *and their reductions* $\overline{f_n} \in \overline{E}$ *form an* $\mathbb{F}_q$-*basis of* $\overline{E}$ *in the algebraic sense, that is the sequence* $\{\overline{f_n}\}$ *is linearly independent, and every element from* $\overline{E}$ *is a finite linear combination of elements of this sequence.*

Proof. The necessity is evident. Let us prove the sufficiency. If $u \in E_0$, denote by $\overline{u}$ its image in $\overline{E}$. Then $\overline{u}$ can be written as a finite linear combination, $\overline{u} = \sum \xi_i \overline{f}_i$, $\xi_i \in \mathbb{F}_q$. Considering ξ_i as classes from O/P, we can take their representatives $z_i^{(1)} \in O$. Then $u - \sum z_i^{(1)} f_i \in PE_0$, so that

$$u = \sum z_i^{(1)} f_i + xz^{(1)}, \quad z^{(1)} \in E_0.$$

Repeating the above procedure for $z^{(1)}$ and iterating we come to a representation

$$u = \sum z_i f_i, \quad z_i \in O, \ z_i \to 0.$$

If $\|u\| = 1$, then necessarily $\sup_i |z_i| = 1$ (otherwise $|z_i| \leq q^{-1}$ for all i, so that $\|u\| < 1$). Since $\|E\| = |K|$, for any $u \in E$ we have $\|u\| = q^n$, $n \in \mathbb{Z}$, so that $\|x^n u\| = 1$, $x^n u = \sum_{i=0}^\infty x^n z_i f_i$, and we have seen that $\sup_i |x^n z_i| = 1$, whence $\sup_i |z_i| = q^n = \|u\|$. ∎

In fact, every separable Banach space E over the field K, such that $\|E\| = |K|$, possesses an orthonormal basis (the Monna-Fleischer theorem, see [93, 90, 98]). In this book we are interested mostly in the explicit construction of orthonormal bases for some function spaces.

Let E_1 and E_2 be Banach spaces over K with the norms $\|\cdot\|_1$ and $\|\cdot\|_2$, respectively. The *topological tensor product* $E_1 \hat{\otimes} E_2$ is the completion of the algebraic tensor product the $E_1 \otimes_K E_2$ with respect to the norm

$$\|u\|_\otimes = \inf \left\{ \max_{1 \leq i \leq r} \|v_i\|_1 \cdot \|w_i\|_2 : \ u = \sum_{i=1}^r v_i \otimes w_i, \ v_i \in E_1, \ w_i \in E_2 \right\} \tag{1.7}$$

where the infimum is taken over all possible representations $u = \sum_{i=1}^r v_i \otimes w_i$,

$v_i \in E_1$, $w_i \in E_2$. See [100] for the proof of the fact that the expression (1.7) indeed defines a norm, and that $\|v \otimes w\|_\otimes = \|v\|_1 \cdot \|w\|_2$. The topological tensor product has the usual functorial properties with respect to continuous linear mappings.

A typical example of a topological tensor product is the Banach space $C(O, V)$ of continuous functions on O with values from a Banach space V (over the field K), with the norm $\|u\| = \sup_{z \in O} \|u(z)\|_V$. For this space, $C(O, V) = C(O, K) \hat{\otimes} V$ (see [100]). If $C_0(O, V)$ is the subspace of $\mathbb{F}_q$-linear continuous functions, then $C_0(O, V) = C_0(O, K) \hat{\otimes} V$. This will follow (see Section 1.2 below) from an explicit uniformly convergent expansion

$$u(z) = \sum_{n=0}^{\infty} c_n f_n(z), \quad c_n \in V, \ z \in O,$$

of an arbitrary function $u \in C_0(O, V)$ in a system of K-valued normalized additive Carlitz polynomials.

More generally, let E_1, E_2 be Banach spaces over K.

Proposition 1.6 *If E_2 has an orthonormal basis $\{f_n\}$, then $E_1 \hat{\otimes} E_2$ is isometrically isomorphic to the space Z of all sequences $Y = \{y_i \in E_1, \|y_i\|_1 \to 0, \text{ as } i \to \infty\}$, with the norm $\|Y\|_\infty = \sup_i \|y_i\|_1$, $Y = \{y_i\}$. The isomorphism is implemented by the mapping*

$$Z \ni Y \mapsto \sum_{i=0}^{\infty} y_i \otimes f_i \in E_1 \hat{\otimes} E_2. \tag{1.8}$$

Proof. It is clear that the series in (1.8) converges in $E_1 \hat{\otimes} E_2$. Denote by π the mapping (1.8). We have

$$\|\pi(Y)\|_\otimes \leq \|Y\|_\infty. \tag{1.9}$$

Let $t = \sum_j \xi_j \otimes \eta_j \in E_1 \otimes_K E_2$ (thus the sum is finite). We can expand η_j:

$$\eta_j = \sum_{i=0}^{\infty} c_{ij} f_i, \quad c_{ij} \in K, \ |c_{ij}| \to 0 \text{ for each } j,$$

so that $\|\eta_j\|_2 = \max_i |c_{ij}|$. Then

$$t = \sum_{i=0}^{\infty} \left(\sum_j c_{ij} \xi_j \right) \otimes f_i,$$

and we have a linear mapping $\omega : E_1 \otimes_K E_2 \to Z$, $\omega(t) = \left\{ \sum_j c_{ij} \xi_j \right\}_i$. Note that $\omega(t)$ does not depend on the representation $t = \sum_j \xi_j \otimes \eta_j$ – if $\sum_{i=0}^{\infty} y_i \otimes f_i = 0$, where $y_i \in E_1$, $\|y_i\|_1 \to 0$, then $y_i = 0$ for all i (one can apply the linear mapping $\lambda \otimes \mathrm{id}$ where $\lambda : E_1 \to K$ is an arbitrary linear functional, and use the basis property of $\{f_i\}$). Now

$$\|\omega(t)\|_\infty = \sup_i \left\| \sum_j c_{ij} \xi_j \right\|_1 \le \sup_i \sup_j |c_{ij}| \cdot \|\xi_j\|_1 = \sup_j \|\xi_j\|_1 \cdot \|\eta_j\|_2.$$

The left-hand side does not depend on the representation of t, thus

$$\|\omega(t)\|_\infty \le \inf \sup_j \|\xi_j\|_1 \cdot \|\eta_j\|_2 = \|t\|_\otimes.$$

This means that ω can be extended by continuity onto $E_1 \hat{\otimes} E_2$, and

$$\|\omega(t)\|_\infty \le \|t\|_\otimes. \tag{1.10}$$

It remains to note that ω and π are inverse to each other, so that the inequalities (1.9) and (1.10) mean that ω and π are both isometric and implement the required isomorphism. ∎

The field $\overline{K}_c$ is obviously a Banach space over K. For this case, Proposition 1.6 means, in particular, that an orthonormal basis $\{f_n\}$ in $C(O, K)$ or $C_0(O, K)$ is simultaneously an orthonormal basis in $C(O, \overline{K}_c)$ and $C_0(O, \overline{K}_c)$ respectively, that is any function $u \in C(O, \overline{K}_c)$ (or $C_0(O, \overline{K}_c)$) admits a uniformly convergent expansion

$$u(t) = \sum_{n=0}^{\infty} c_n f_n(t), \quad t \in O,$$

where $c_n \in \overline{K}_c$, $|c_n| \to 0$, as $n \to \infty$, and

$$\sup_{t \in O} |u(t)| = \max_{n \ge 0} |c_n|.$$

1.2 Additive Carlitz polynomials

1.2.1. A basis in $C_0(O,K)$. The system of *additive* (in fact, $\mathbb{F}_q$-linear) *Carlitz polynomials* is defined as follows. Let $e_0(t) = t$,

$$e_i(t) = \prod_{\substack{m \in \mathbb{F}_q[x] \\ \deg m < i}} (t - m), \quad i \geq 1. \tag{1.11}$$

By Proposition 1.2, the polynomials e_i are $\mathbb{F}_q$-linear.

Proposition 1.7 (i) *The polynomials e_i satisfy the recursive relations*

$$e_i(t) = e_{i-1}^q(t) - D_{i-1}^{q-1} e_{i-1}(t), \quad i \geq 1; \tag{1.12}$$

$$e_i(xt) = xe_i(t) + [i]e_{i-1}^q(t), \quad i \geq 1. \tag{1.13}$$

(ii) *There is an explicit representation*

$$e_i(t) = \sum_{j=0}^{i} (-1)^{i-j} \begin{bmatrix} i \\ j \end{bmatrix} t^{q^j} \tag{1.14}$$

where

$$\begin{bmatrix} i \\ j \end{bmatrix} = \frac{D_i}{D_j L_{i-j}^{q^j}}.$$

(iii) *The relations for special values*

$$e_i(x^i) = D_i, \tag{1.15}$$

$$e_i\left(\frac{1}{x+1}\right) = \frac{(-1)^i D_i}{(x+1)^l}, \quad l = 1 + q + \cdots + q^i, \tag{1.16}$$

hold for any $i = 0, 1, 2, \ldots$

Proof. First we prove (1.15). The class of polynomials $x^i - m$, $m \in \mathbb{F}_q[x]$, $\deg m < i$, is exactly the class of all monic polynomials of degree i. Therefore (1.15) follows from Lemma 1.1.

In order to prove (1.12), note that both sides are monic polynomials of the same degree q^i. Thus we need only to show that they have the same set of roots $m \in \mathbb{F}_q[x]$, $\deg m < i$ (the number of such polynomials is just q^i). It follows from the definition (1.11) that both sides of (1.12) vanish on $t \in \mathbb{F}_q[x]$, $\deg t < i-1$. If $t = \zeta x^{i-1}$, $\zeta \in \mathbb{F}_q$, then $e_i(t) = 0$, while the right-hand side of (1.12) equals $\zeta \left(D_{i-1}^q - D_{i-1}^{q-1} \cdot D_{i-1} \right) = 0$, which proves (1.12).

The identities (1.14) and (1.16) are obvious for $i = 0$ and follow, for any i, by induction from (1.12). Using (1.14) we can check (1.13) by a direct computation. ∎

Denote by f_i the normalized *Carlitz polynomials*:

$$f_i(t) = \frac{e_i(t)}{D_i}, \quad i = 0, 1, 2, \ldots$$

Let us introduce also a sequence of difference operators which will play a crucial role throughout this book. Let

$$(\Delta u)(t) = u(xt) - xu(t), \tag{1.17}$$

$$\left(\Delta^{(j)} u \right)(t) = \left(\Delta^{(j-1)} u \right)(xt) - x^{q^{j-1}} u(t), \quad j \geq 1, \tag{1.18}$$

where $\Delta^{(0)} = I$ (the identity operator), so that $\Delta^{(1)} = \Delta$.

It is easy to prove by induction that

$$\left(\Delta^{(j)} u \right)(t) = u(x^j t)$$
$$+ \sum_{k=0}^{j-1} (-1)^{j-k} \left(\sum_{0 \leq i_1 < \ldots < i_{j-k} \leq j-1} x^{q^{i_1} + \cdots + q^{i_{j-k}}} \right) u(x^k t). \tag{1.19}$$

The identity (1.13) can be written as

$$(\Delta e_i)(t) = [i] e_{i-1}^q(t), \quad i \geq 1, \tag{1.20}$$

whence

$$(\Delta f_i)(t) = f_{i-1}^q(t), \quad i \geq 1. \tag{1.21}$$

Obviously, $\Delta e_0 = \Delta f_0 = 0$. More generally, we prove by induction that

$$\Delta^{(j)} e_k = \begin{cases} \dfrac{D_k}{D_{k-j}^{q^j}} e_{k-j}^{q^j}, & \text{if } k \geq j, \\ 0, & \text{if } k < j, \end{cases} \tag{1.22}$$

so that

$$\Delta^{(j)} f_k = \begin{cases} f_{k-j}^{q^j}, & \text{if } k \geq j, \\ 0, & \text{if } k < j. \end{cases} \tag{1.23}$$

Theorem 1.8 *The sequence $\{f_i\}_0^\infty$ is an orthonormal basis of the Banach space $C_0(O, K)$ of $\mathbb{F}_q$-linear K-valued continuous functions on the ring of integers $O \subset K$. The coefficients of the expansion*

$$u(t) = \sum_{i=0}^{\infty} c_i f_i(t), \quad t \in O, \quad c_i \to 0, \tag{1.24}$$

of an arbitrary function $u \in C_0(O, K)$ can be written explicitly:

$$c_i = \Delta^{(i)} u(1), \quad i = 0, 1, 2, \ldots \tag{1.25}$$

Proof. It follows from (1.21) that

$$f_i(x^{n+1}) = x f_i(x^n) + f_{i-1}^q(x^n), \quad n \geq 0.$$

By induction and $\mathbb{F}_q$-linearity, we see that $f_i(t) \in \mathbb{F}_q[x]$ for any $t \in \mathbb{F}_q[x]$. Since $\mathbb{F}_q[x]$ is dense in O, we get $\|f_i\| \leq 1$ (the supremum norm), and the identity (1.15) implies the equalities $\|f_i\| = 1$, $i = 0, 1, 2, \ldots$

If u is $\mathbb{F}_q$-linear, we have $u(0) = u(pt) = pu(t) = 0$, and if u is also continuous, then $u(x^i) \to 0$, as $i \to \infty$. From (1.19) we find for the coefficients (1.25) the estimate

$$|c_i| \leq \max\left\{|u(x^i)|, \max_{0 \leq k \leq i-1} q^{k-i} |u(x^k)|\right\}$$

$$\leq \max\left\{|u(x^i)|, \|u\| q^{-i/2}, \max_{k > i/2} |u(x^k)|\right\} \longrightarrow 0,$$

as $i \to \infty$. Therefore the series in (1.24), with the coefficients given by (1.25), converges in $C_0(O, K)$. Let us prove that it converges to u.

Denote the right-hand side of (1.24) temporarily by $v(t)$, so that

$$v(t) = \sum_{i=0}^{\infty} c_i f_i(t), \quad t \in O. \tag{1.26}$$

Applying a linear bounded operator $\Delta^{(k)}$ to both sides of (1.26) we find that

$$\left(\Delta^{(k)} v\right)(t) = \sum_{i=k}^{\infty} c_i f_{i-k}^{q^k}(t),$$

so that $c_i = \Delta^{(i)} v(1)$, and by (1.25)

$$\Delta^{(i)}(u - v)(1) = 0, \quad i = 0, 1, 2, \ldots$$

Using (1.19) we find consequently that $u(x^n) = v(x^n)$ for each n. Therefore, by $\mathbb{F}_q$-linearity, u and v coincide on $\mathbb{F}_q[x]$ and, by continuity, $u = v$, and we have proved (1.24). This equality shows that $\|u\| \leq \max_i |c_i|$. The opposite inequality follows from (1.25). ∎

Remarks 1.1 (1) Theorem 1.8 and its proof remain valid for functions with values from a Banach space V over K. As mentioned in Section 1.1.3, this proves the equality $C_0(O, V) = C_0(O, K) \hat{\otimes} V$.

(2) The condition $c_i \to 0$ is also necessary for the convergence of the series (1.24) on O. Indeed, take $t = (1+x)^{-1}$. By (1.16),

$$\left| f_i \left(\frac{1}{x+1} \right) \right| = 1,$$

and the convergence of (1.24) at this value of t means that $c_i \to 0$, as $i \to \infty$.

(3) If m is a fixed positive integer, then $\left\{ f_i^{q^m} \right\}_{i=0}^\infty$ is an orthonormal basis in $C_0(O, K)$. Actually, this is a very general fact, valid for bases in any spaces of K-valued continuous functions [53]. For the proof note that the difference $f_i^{q^m} - f_i$ takes its values in the maximal ideal P of O. Therefore $\|f_i^{q^m} - f_i\| \leq q^{-1}$ for all values of i, and the result is a consequence of Proposition 1.5 – the reductions of both sequences (considered in Proposition 1.5) are the same.

(4) It follows from Theorem 1.8 and the arguments used in its proof that the sequence $\{f_i\}_0^\infty \subset \mathbb{A}[t]$, $\mathbb{A} = \mathbb{F}_q[x]$, forms an $\mathbb{A}$-basis (in the algebraic sense) of an $\mathbb{A}$-module $\mathrm{Int}_O(\mathbb{A})$ of $\mathbb{F}_q$-linear integer-valued polynomials from $\mathbb{A}[t]$ (a polynomial is called *integer-valued* if $f(t) \in \mathbb{A}$ for any $t \in \mathbb{A}$). The expression (1.25) for the coefficients of expansions in the Carlitz polynomials remains valid in this situation.

1.2.2. Strongly singular functions. The series (1.24) always makes sense for $t \in \mathbb{F}_q[x]$ – for each such t only a finite number of terms is different from zero. A function $u(t)$, $t \in \mathbb{F}_q[x]$, defined by the series (1.24), is called *strongly singular*, if the series does not converge for any element $t \in O \setminus \mathbb{F}_q[x]$.

We will need the following property of the Carlitz polynomials. As we know, $f_i(x^n) = 0$, if $n < i$. Let us consider the case where $n \geq i$.

Lemma 1.9 *If $n \geq i$, then $|f_i(x^n)| = q^{i-n}$.*

Proof. For any $\omega \in \mathbb{F}_q[x]$, $\deg \omega < i$, we have $|x^n - \omega| = |\omega|$. Writing

$$e_i(t) = t \prod_{\substack{0 \neq \omega \in \mathbb{F}_q[x] \\ \deg \omega < i}} (t - \omega)$$

we find that

$$|e_i(x^n)| = |x^n| \prod_{\substack{\deg \omega < i \\ \omega \neq 0}} |\omega| = |x^n| \cdot \left|\lim_{t \to 0} \frac{e_i(t)}{t}\right|.$$

It follows from (1.14) that

$$\lim_{t \to 0} \frac{e_i(t)}{t} = (-1)^i \frac{D_i}{L_i}$$

whence

$$|f_i(x^n)| = \frac{q^{-n}}{|L_i|} = q^{i-n}$$

as desired. ∎

Now we get a general sufficient condition for a function (1.24) to be strongly singular.

Theorem 1.10 *If $|c_i| \geq \rho > 0$ for all $i \geq i_0$ (where i_0 is some natural number), then the function (1.24) is strongly singular.*

Proof. It is sufficient to find, for any $t \in O \setminus \mathbb{F}_q[x]$, a sequence $i_k \to \infty$ such that $|f_{i_k}(t)| = 1$, $k = 1, 2, \ldots$.

In fact, if $t \in O \setminus \mathbb{F}_q[x]$, then $t = \sum_{n=0}^{\infty} \xi_n x^n$, $\xi_n \in \mathbb{F}_q$, with $\xi_{i_k} \neq 0$ for some sequence $i_k \to \infty$. Then

$$f_{i_k}(t) = \sum_{n=i_k}^{\infty} \xi_n f_{i_k}(x^n)$$

where $|f_{i_k}(x^{i_k})| = 1$ by (1.15), $|f_{i_k}(x^n)| = q^{i_k - n} \leq q^{-1}$ for $n > i_k$ (by Lemma 1.9), and $|\xi_{i_k}| = 1$. Therefore $|f_{i_k}(t)| = 1$ for any k. ∎

1.3 Hyperdifferentiations

1.3.1. Definitions and main identities. *Hyperdifferentiations* (or the

Hasse derivatives) form a sequence of K-valued functions $\mathcal{D}_k(t)$, $k \geq 0$, $t \in O$, defined as follows. Set $\mathcal{D}_0(x^n) = x^n$, $\mathcal{D}_k(1) = 0$ for $k \geq 1$,

$$\mathcal{D}_k(x^n) = \binom{n}{k} x^{n-k}, \tag{1.27}$$

where it is assumed that $\binom{n}{k} = 0$ for $k > n$. The binomial coefficients are natural numbers making sense as elements of K. They can be defined by the recurrence relation

$$\binom{n}{k} = \binom{n-1}{k} + \binom{n-1}{k-1} \tag{1.28}$$

with appropriate boundary conditions; this definition does not involve divisions which are impossible in positive characteristic.

Then $\mathcal{D}_k$ is extended onto $\mathbb{F}_q[x]$ by $\mathbb{F}_q$-linearity:

$$\mathcal{D}_k \left(\sum_{n=n_0}^{N} \xi_n x^n \right) = \sum_{n=\max(n_0,k)}^{N} \binom{n}{k} \xi_n x^{n-k}, \quad \xi_n \in \mathbb{F}_q, \xi_{n_0} \neq 0,$$

$n_0 \in \mathbb{Z}_+$, so that

$$\left| \mathcal{D}_k \left(\sum_{n=n_0}^{N} \xi_n x^n \right) \right| \leq q^{-(\max(n_0,k)-k)} \leq q^k \left| \sum_{n=n_0}^{N} \xi_n x^n \right|.$$

Therefore each function $\mathcal{D}_k$ is continuous at the origin (thus on $\mathbb{F}_q[x]$, due to additivity), and can be extended to an $\mathbb{F}_q$-linear continuous function on O.

Note that each function $\mathcal{D}_k$ ($k \geq 1$) is nowhere differentiable. Indeed, it is sufficient to check that $\mathcal{D}_k$ is not differentiable at the origin, that is to find a sequence $t_n \in O$ such that $t_n \to 0$, as $n \to \infty$, but $t_n^{-1} \mathcal{D}_k(t_n)$ does not converge. Set $t_n = x^n$; then $t_n \to 0$, but $t_n^{-1} \mathcal{D}_k(t_n) = \binom{n}{k} x^{-k}$ does not converge.

The definition (1.27) shows that in characteristic zero one would have $\mathcal{D}_k = \frac{1}{k!} \frac{d^k}{dx^k}$, but this expression does not make sense in positive characteristic. However the hyperdifferentiations share some properties with the usual derivatives. One has the Leibniz rule

$$\mathcal{D}_k(st) = \sum_{j=0}^{k} D_j(s) D_{k-j}(t), \quad s, t \in O. \tag{1.29}$$

Indeed, by continuity and $\mathbb{F}_q$-linearity, the proof of (1.29) reduces to the case $s = x^m$, $t = x^n$, where the Leibnitz rule reduces to the Vandermonde formula

$$\binom{m+n}{k} = \sum_{j=0}^{k} \binom{m}{k}\binom{n}{k-j} \qquad (1.30)$$

(see [88]).

On the other hand, the form of the sequence $\mathcal{D}_k$ (that is, the expression (1.27)) follows from (1.29) and the assumptions $\mathcal{D}_0(x) = x$, $\mathcal{D}_1(x) = 1$, $\mathcal{D}_k(x) = 0$ for $k \geq 2$.

The relations between hyperdifferentiations and polynomial systems are given by the following results.

Proposition 1.11 (i) *For any $t \in O$,*

$$\mathcal{D}_k(t) = \sum_{n=k}^{\infty} A_{nk} f_n(t) \qquad (1.31)$$

where f_n are the normalized additive Carlitz polynomials, $A_{n1} = (-1)^{n-1} L_{n-1}$,

$$A_{nk} = (-1)^{n+k} L_{n-1} \sum_{0 < i_1 < \ldots < i_{k-1} < n} \frac{1}{[i_1][i_2]\ldots[i_{k-1}]}, \quad k > 1. \qquad (1.32)$$

(ii) *For any $k \geq 0$, we have $\|\mathcal{D}_k\| = 1$ (as before, $\|\cdot\|$ is the norm in $C_0(O, K)$).*

(iii) *For any $t \in O$, $n \in \mathbb{N}$,*

$$t^{q^n} = \sum_{k=0}^{\infty} [n]^k \mathcal{D}_k(t). \qquad (1.33)$$

(iv) *For any $t \in O$,*

$$f_n(t) = \sum_{k=n}^{\infty} B_{kn} \mathcal{D}_k(t), \quad n = 0, 1, 2, \ldots, \qquad (1.34)$$

where

$$B_{kn} = \sum_{i=0}^{n} \frac{(-1)^{n-i}}{D_i L_{n-i}^{q^i}} [i]^k.$$

Proof. (i) By Theorem 1.8, we have to show that $A_{nk} = \Delta^{(n)}\mathcal{D}_k(1)$. In fact, we will prove that

$$\Delta^{(n)}\mathcal{D}_k(t) = \sum_{i=0}^{k-1} A_{n,k-i}\mathcal{D}_i(t), \quad t \in O, \ k \geq 1, \tag{1.35}$$

which gives the required identity for $t = 1$.

We proceed by induction on n. It is checked directly, using (1.27) and (1.28), that

$$\Delta\mathcal{D}_k = \mathcal{D}_{k-1}, \ k \geq 1; \quad \Delta\mathcal{D}_0 = 0. \tag{1.36}$$

If (1.35) holds for some n, then, writing $\mathcal{D}_{-1} = 0, A_{n0} = 0$, we get from the recursive definition (1.18) that

$$\Delta^{(n+1)}\mathcal{D}_k(t) = \sum_{i=0}^{k-1} \left\{ A_{n,k-i}\mathcal{D}_i(xt) - x^{q^n} A_{n,k-i}\mathcal{D}_i(t) \right\}$$

$$= \sum_{i=0}^{k-1} \left\{ A_{n,k-i} x \mathcal{D}_i(t) - A_{n,k-i}\mathcal{D}_{i-1}(t) - x^{q^n} A_{n,k-i}\mathcal{D}_i(t) \right\}$$

$$= \sum_{i=0}^{k-1} \left\{ -A_{n,k-i}[n] + A_{n,k-i-1} \right\} \mathcal{D}_i(t) = \sum_{i=0}^{k-1} A_{n+1,k-i}\mathcal{D}_i(t),$$

which completes the proof.

(ii) It is clear that $\|\mathcal{D}_0\| = 1$. For $\mathcal{D}_1$ we find that $\|\mathcal{D}_1\| = \max_{n \geq 1} \left\{ q^{-n+1} \right\} = 1$. Next, by (1.32),

$$A_{nk} \leq q^{-n+1} \max_{0 < i_1 < \ldots < i_{k-1} < n} q^{k-1} = q^{-n+k},$$

so that $|A_{nk}| < 1$, if $k < n$. As we know, $|A_{nn}| = 1$. The required result follows from the fact that $\|f_n\| = 1$ for any n.

(iii) By (ii), the series in (1.33) is uniformly convergent, and it is sufficient to check the equality (1.33) for $t = x^m$, $m \in \mathbb{N}$. For this case, (1.33) follows easily from the Newton binomial formula.

(iv) We find from (1.14) and (1.33) that

$$f_n(t) = \sum_{k=0}^{\infty} B_{kn}\mathcal{D}_k(t), \quad t \in O, \ n = 0, 1, 2, \ldots, \tag{1.37}$$

where B_{kn} has the required form. In (1.37) we set successively $t = 1, x, \ldots, x^{n-1}, x^n$. Since $f_n(x^n) = 1$ and $f_n(x^j) = 0$ for $j < n$, we find that $B_{kn} = 0$ for $k < n$. Thus, (1.37) gives the required identity (1.34). ∎

Orthonormal systems

1.3.2. The basis property. The next result, the main one in this section, shows not only that $\{\mathcal{D}_k\}$ is an orthonormal basis, but also that the sequence of hyperdifferentiations is closely connected with powers Δ^n of the Carlitz difference operator, just as the sequence $\{f_k\}$ of normalized Carlitz polynomials is connected with higher difference operators $\Delta^{(n)}$.

Theorem 1.12 *The sequence $\{\mathcal{D}_k\}_0^\infty$ is an orthonormal basis of the Banach space $C_0(O, K)$. The coefficients of the expansion*

$$u(t) = \sum_{k=0}^\infty c_k \mathcal{D}_k(t), \quad t \in O, \tag{1.38}$$

of an arbitrary function $u \in C_0(O, K)$ can be written explicitly:

$$c_k = \Delta^k u(1) = \sum_{i=0}^n (-1)^{n-i} u(x^i) \mathcal{D}_i(x^n), \quad k = 0, 1, 2, \ldots \tag{1.39}$$

Proof. Using the representation (1.31), in which $A_{nn} = 1$ and $|A_{nk}| \leq q^{-1}$ for $k < n$ (see the Proof of Proposition 1.11) we find that

$$\|f_k - \mathcal{D}_k\| \leq q^{-1}, \quad k = 0, 1, 2, \ldots$$

Since $\{f_k\}$ is an orthonormal basis of $C_0(O, K)$, it follows from Proposition 1.5 that the sequence $\{\mathcal{D}_k\}$ possesses the same property.

In order to prove the first equality in (1.39), it suffices to apply the operator Δ to the equality (1.38), to take into account the identity (1.36), to set $t = 1$, then to repeat the procedure, etc.

To get the second equality, we use induction on n. For $n = 0$, the required identity is evident, since $\mathcal{D}_0(1) = 1$ and $c_0 = u(1)$. Suppose it holds for $k < n$. Let us take $t = x^n$ in (1.38). As $\mathcal{D}_n(x^n) = 1$, $\mathcal{D}_k(x^n) = 0$ for $k > n$, we get

$$u(x^n) = c_n + \sum_{k=0}^{n-1} c_k \mathcal{D}_k(x^n),$$

so that

$$c_n = u(x^n) \mathcal{D}_n(x^n) - \sum_{k=0}^{n-1} c_k \mathcal{D}_k(x^n).$$

The induction hypothesis allows us to rewrite this as

$$c_n = u(x^n)\mathcal{D}_n(x^n) - \sum_{k=0}^{n-1}\left(\sum_{j=0}^{k}(-1)^{k-j}u(x^j)\mathcal{D}_j(x^k)\right)\mathcal{D}_k(x^n)$$

$$= u(x^n)\mathcal{D}_n(x^n) - \sum_{j=0}^{n-1}\left(\sum_{k=j}^{n-1}(-1)^{k-j}\mathcal{D}_j(x^k)\mathcal{D}_k(x^n)\right)u(x^j)$$

$$= u(x^n)\mathcal{D}_n(x^n) - \sum_{j=0}^{n-1}\left(\sum_{k=j}^{n-1}(-1)^{k-j}\binom{k}{j}\binom{n}{k}x^{n-j}\right)u(x^j).$$

Using simple binomial identities [88]

$$\binom{k}{j}\binom{n}{k} = \binom{n}{j}\binom{n-j}{k-j}, \tag{1.40}$$

$$\sum_{k=j}^{n-1}(-1)^{k-j}\binom{n-j}{k-j} = (-1)^{n-j+1},$$

we obtain that

$$c_n = u(x^n)\mathcal{D}_n(x^n) + \sum_{j=0}^{n-1}(-1)^{n-j}\binom{n}{j}x^{n-j}u(x^j),$$

and this is just the required formula. ∎

As we discussed in Section 1.1.2, typical classes of $\mathbb{F}_q$-linear functions form rings with composition as the multiplication operation. In connection with this, it is interesting to note the following property of hyperdifferentiations:

$$\mathcal{D}_k(\mathcal{D}_l(t)) = \binom{k+l}{l}\mathcal{D}_{k+l}(t), \quad t \in O.$$

Its proof follows from the identity (1.40) and the definitions.

1.4 The digit principle

1.4.1. A general theorem. This section is devoted to the following problems. Given an orthonormal basis $\{\varphi_j\}_0^\infty$ of the space $C_0(O, K)$ of $\mathbb{F}_q$-linear continuous functions, how do we construct an orthonormal basis $\{\Phi_j\}_0^\infty$ of the space $C(O, K)$ of all continuous functions? How do we find the coefficients of the expansions in such bases? While the second problem is connected with special properties of each basis, the first one admits a general solution.

Let us write every integer $i \geq 0$ in the base q as

$$i = \alpha_0 + \alpha_1 q + \cdots + \alpha_{n-1} q^{n-1}, \quad 0 \leq \alpha_j \leq q-1.$$

Set

$$\Phi_i = \varphi_0^{\alpha_0} \varphi_1^{\alpha_1} \cdots \varphi_{n-1}^{\alpha_{n-1}} \tag{1.41}$$

where it is assumed that $\varphi_j^\alpha = 1$ for $\alpha = 0$ even if φ_j vanishes at some points. Note that $\varphi_j = \Phi_{q^j}$. The construction (1.41) is called the extension of the basis $\{\varphi_j\}$ by digit expansions.

Theorem 1.13 ("Digit principle") *The extension of an orthonormal basis of $C_0(O, K)$ by digit expansions produces an orthonormal basis for $C(O, K)$.*

Proof. Using Proposition 1.5 we reduce our problem to that of linear algebra over the finite field $\mathbb{F}_q$. The reduction of the space $C_0(O, K)$ is $\mathrm{Hom}_{\mathbb{F}_q}(O, \mathbb{F}_q) = \bigoplus_{j \geq 0} \mathbb{F}_q \overline{\varphi_j}$ where the reductions $\{\overline{\varphi_j}\}_0^\infty$ form an $\mathbb{F}_q$-basis of $\mathrm{Hom}_{\mathbb{F}_q}(O, \mathbb{F}_q)$. Here $\mathrm{Hom}_{\mathbb{F}_q}(O, \mathbb{F}_q)$ denotes the set of all continuous $\mathbb{F}_q$-linear maps (with respect to the discrete topology in $\mathbb{F}_q$). Similarly we will denote by $C(O, \mathbb{F}_q)$ the set of all continuous functions $O \to \mathbb{F}_q$, the reduction of the Banach space $C(O, K)$.

Let $H_n = \bigcap_{j=0}^{n-1} \ker(\overline{\varphi_j})$, $n \geq 1$. Then H_n is a closed subspace of $\mathbb{F}_q$-codimension n in O, $H_{n+1} \subset H_n$, and $\bigcap_{n=1}^{\infty} H_n = \{0\}$. Therefore $O \cong \varprojlim O/H_n$. Note that O/H_n is a finite set for each n. Denoting by $\mathrm{Maps}(X, Y)$ the set of all mappings from a finite set X to a finite set Y, we see that

$$C(O, \mathbb{F}_q) = \varinjlim \mathrm{Maps}(O/H_n, \mathbb{F}_q).$$

Considering $\overline{\varphi_0}, \ldots, \overline{\varphi_{n-1}}$ as functions on O/H_n we see that they form an $\mathbb{F}_q$-basis of the $\mathbb{F}_q$-dual space $(O/H_n)^*$. Thus our problem is reduced to the following one.

Let V be a finite-dimensional $\mathbb{F}_q$-vector space, $\dim V = n$, and let $\psi_0, \ldots, \psi_{n-1}$ be a basis of V^*. Extend the functionals ψ_j to a system of polynomial functions on V by using digit expansions, that is for $0 \leq i \leq q^n - 1$ write $i = \alpha_0 + \alpha_1 q + \cdots + \alpha_{n-1} q^{n-1}$ and set

$$\Psi_i = \psi_0^{\alpha_0} \cdots \psi_{n-1}^{\alpha_{n-1}},$$

so that $\psi_j = \Psi_{q^j}$ and $\Psi_0 = 1$. We need to show that the functions Ψ_i form a basis of $\mathrm{Maps}(V, \mathbb{F}_q)$. By a dimension count, it suffices to prove that the Ψ_i span $\mathrm{Maps}(V, \mathbb{F}_q)$.

Let $\{v_0, \ldots, v_{n-1}\}$ be the dual basis to $\{\psi_0, \ldots, \psi_{n-1}\}$. For $v \in V$, write

$$v = a_0 v_0 + a_1 v_1 + \cdots + a_{n-1} v_{n-1}$$

where $a_j \in \mathbb{F}_q$. Consider functions $h_v : V \to \mathbb{F}_q$ of the form

$$h_v(w) = \prod_{j=0}^{n-1} \left(1 - (\psi_j(w) - a_j)^{q-1}\right) = \prod_{j=0}^{n-1} \left(1 - (\psi_j(w) - \psi_j(v))^{q-1}\right).$$

Since $h_v(w) = 1$ for $v = w$ and $h_v(w) = 0$ when $w \neq v$ (because at least one of the differences $\psi_j(w) - \psi_j(v)$ is in this case a nonzero element of the finite field $\mathbb{F}_q$), the $\mathbb{F}_q$-span of all the h_v is the whole set $\mathrm{Maps}(V, \mathbb{F}_q)$. Expanding the product defining h_v shows that h_v is in the span of the Ψ_i since the exponents of the ψ_j in the product never exceed $q - 1$. This completes the proof. ∎

1.4.2. The general Carlitz polynomials. Let us apply the construction (1.41) to the case where $\varphi_j = f_j$, the normalized Carlitz polynomials. Denote by $\{G_i\}$ the resulting polynomial system, so that $G_0(t) = 1$,

$$G_i(t) = \prod_{j=0}^{n-1} f_j(t)^{\alpha_j} = \frac{1}{\Gamma_i} \prod_{j=0}^{n-1} e_j(t)^{\alpha_j}$$

where the factorial-like sequence $\{\Gamma_i\}$ is defined by (1.4). By Theorem 1.13, $\{G_i\}$ is an orthonormal basis of $C(O, K)$.

Let us study expansions with respect to this basis in detail. We will need an auxiliary polynomial system

$$g_i(t) = \prod_{j=0}^{n-1} g_{\alpha_j q^j}(t)$$

where

$$g_{\alpha_j q^j}(t) = \begin{cases} f_j^{\alpha_j}(t), & \text{if } 0 \leq \alpha_j < q - 1; \\ f_j^{\alpha_j}(t) - 1, & \text{if } \alpha_j = q - 1. \end{cases}$$

Proposition 1.14 *The polynomial systems $\{G_i\}$ and $\{g_i\}$ have the following properties:*

(i) $G_i(\xi t) = \xi^i G_i(t)$, $\xi \in \mathbb{F}_q$; $g_i(\xi t) = \xi^i g_i(t)$, $0 \neq \xi \in \mathbb{F}_q$.   (1.42)

(ii) $G_i(t+s) = \sum_{j+l=i} \binom{i}{j} G_j(t) G_l(s);$ (1.43)

$$g_i(t+s) = \sum_{j+l=i} \binom{i}{j} G_j(t) g_l(s) = \sum_{j+l=i} \binom{i}{j} g_j(t) G_l(s). \quad (1.44)$$

(iii) *If $0 \leq l < q^\nu$, $k \geq 0$, $\nu \in \mathbb{N}$, then*

$$\sum_{\substack{m \in \mathbb{F}_q[x] \\ \deg m < \nu}} G_k(m) g_l(m) = \begin{cases} 0, & \text{if } k+l \neq q^\nu - 1, \\ (-1)^\nu, & \text{if } k+l = q^\nu - 1. \end{cases} \quad (1.45)$$

(iv) *If $0 \leq k, l < q^\nu$, then*

$$\sum_{\substack{m \text{ monic} \\ \deg m = \nu}} G_k(m) g_l(m) = \begin{cases} 0, & \text{if } k+l \neq q^\nu - 1, \\ (-1)^\nu, & \text{if } k+l = q^\nu - 1. \end{cases} \quad (1.46)$$

Proof. (i) The first equality in (1.42) follows from the congruence

$$\sum_{j=0}^{n-1} \alpha_j \equiv \sum_{j=0}^{n-1} \alpha_j q^j \pmod{q-1}.$$

Similarly we get the second equality if $\alpha_j < q-1$ for all j. If some $\alpha_j = q-1$, then for that j we get $g_{\alpha_j q^j}(\xi t) = g_{\alpha_j q^j}(t)$, if $\xi \neq 0$. Therefore we come to the required equality assuming that $\xi \neq 0$.

(ii) To prove (1.43), note that

$$f_j(t+s)^{\alpha_j} = (f_j(t) + f_j(s))^{\alpha_j} = \sum_{l=0}^{\alpha_j} \binom{\alpha_j}{l} f_j(t)^l f_j(s)^{\alpha_j - l},$$

whence

$$G_j(t+s) = \prod_{j=0}^{n-1} \sum_{l_j=0}^{\alpha_j} \binom{\alpha_j}{l_j} f_j(t)^{l_j} f_j(s)^{\alpha_j - l_j}$$

$$= \sum_{l_0=0}^{\alpha_0} \cdots \sum_{l_{n-1}=0}^{\alpha_{n-1}} \left\{ \binom{\alpha_0}{l_0} \cdots \binom{\alpha_{n-1}}{l_{n-1}} f_0(t)^{l_0} \cdots f_{n-1}(t)^{l_{n-1}} \right.$$

$$\left. \times f_0(s)^{\alpha_0 - l_0} \cdots f_{n-1}(s)^{\alpha_{n-1} - l_{n-1}} \right\}$$

$$= \sum_{l_0=0}^{\alpha_0} \cdots \sum_{l_{n-1}=0}^{\alpha_{n-1}} \binom{\alpha_0}{l_0} \cdots \binom{\alpha_{n-1}}{l_{n-1}} G_\beta(t) G_{i-\beta}(s),$$

with $\beta = l_0 + l_1 q + \cdots + l_{n-1} q^{n-1}$. It follows from Lucas' theorem about binomial coefficients modulo a prime number (see [46]) that

$$\binom{i}{\beta} \equiv \binom{\alpha_0}{l_0} \cdots \binom{\alpha_{n-1}}{l_{n-1}} \pmod{p}, \qquad (1.47)$$

and it follows from (1.47) that

$$\binom{i}{\beta} \equiv 0 \pmod{p} \quad \text{if } \beta = l_0 + l_1 q + \cdots + l_{n-1} q^{n-1}$$

with $l_j > \alpha_j$ for some value of j. This results in (1.43), as well as in (1.44) for the case where $\alpha_j < q - 1$ for every j.

If $\alpha_j = q - 1$ for some j, then

$$g_{\alpha_j q^j}(t+s) = (f_j(t) + f_j(s))^{q-1} - 1 = \sum_{\beta + l = q - 1} \binom{q-1}{\beta} f_j(t)^\beta f_j(s)^l - 1.$$

Considering separately the terms with $\beta = 0$ and $\beta = q - 1$, and repeating the above reasoning we come to (1.44).

(iii) Writing k in the q-digit expansion, $k = k_0 + k_1 q + \cdots + k_{n-1} q^{n-1}$, $k_{n-1} \neq 0$, we have

$$G_k(t) = f_0(t)^{k_0} f_1(t)^{k_1} \cdots f_{n-1}(t)^{k_{n-1}}.$$

If $m \in \mathbb{F}_q[x]$ with $\deg m < \nu$, then $f_j(m) = 0$ for $j \geq \nu$. Therefore we need only prove (1.45) in the case where $k < q^\nu$.

Thus, let $k = k_0 + k_1 q + \cdots + k_{\nu-1} q^{\nu-1}$, $l = l_0 + l_1 q + \cdots + l_{\nu-1} q^{\nu-1}$, with $0 \leq k_i \leq q - 1$, $0 \leq l_i \leq q - 1$ for $0 \leq i \leq \nu - 1$. We have

$$G_k(t) = f_0(t)^{k_0} f_1(t)^{k_1} \cdots f_{\nu-1}(t)^{k_{\nu-1}},$$

$$g_l(t) = \left(f_0(t)^{l_0} - \delta_0\right)\left(f_1(t)^{l_1} - \delta_1\right)\cdots\left(f_{\nu-1}(t)^{l_{\nu-1}} - \delta_{\nu-1}\right),$$

where $\delta_j = \delta_{q-1,l_j}$ for $0 \leq j \leq \nu - 1$. Therefore

$$G_k(t)g_l(t) = \left(f_0(t)^{k_0+l_0} - \delta_0 f_0(t)^{k_0}\right)\left(f_1(t)^{k_1+l_1} - \delta_1 f_1(t)^{k_1}\right)\cdots \\ \times \left(f_{\nu-1}(t)^{k_{\nu-1}+l_{\nu-1}} - \delta_{\nu-1}f_{\nu-1}(t)^{k_{\nu-1}}\right). \quad (1.48)$$

Denote $H_i^s = f_i(x^{i+s})$, $i, s \geq 0$. Let $m \in \mathbb{F}_q[x]$ with $\deg m < \nu$. Then $m = c_{\nu-1}x^{\nu-1} + \cdots + c_1 x + c_0$, $c_j \in \mathbb{F}_q$, so that

$$f_i(m) = c_{\nu-1}H_i^{\nu-1-i} + \cdots + c_{i+1}H_i^1 + c_i H_i^0, \quad 0 \leq i \leq \nu - 1.$$

Now, by (1.48),

$$\sum_{\deg m < \nu} G_k(m)g_l(m)$$

$$= \sum_{c_0,c_1,\ldots,c_{\nu-1} \in \mathbb{F}_q} \prod_{i=0}^{\nu-1}\left(\left(c_{\nu-1}H_i^{\nu-1-i} + \cdots + c_{i+1}H_i^1 + c_i H_i^0\right)^{k_i+l_i}\right.$$

$$\left. - \delta_i \left(c_{\nu-1}H_i^{\nu-1-i} + \cdots + c_{i+1}H_i^1 + c_i H_i^0\right)^{k_i}\right)$$

$$= \sum_{c_1,\ldots,c_{\nu-1} \in \mathbb{F}_q} \prod_{i=1}^{\nu-1}\left(\left(c_{\nu-1}H_i^{\nu-1-i} + \cdots + c_{i+1}H_i^1 + c_i H_i^0\right)^{k_i+l_i}\right.$$

$$\left. - \delta_i \left(c_{\nu-1}H_i^{\nu-1-i} + \cdots + c_{i+1}H_i^1 + c_i H_i^0\right)^{k_i}\right)$$

$$\times \sum_{c_0 \in \mathbb{F}_q}\left(\left(c_{\nu-1}H_0^{\nu-1} + \cdots + c_1 H_0^1 + c_0 H_0^0\right)^{k_0+l_0}\right.$$

$$\left. - \delta_0 \left(c_{\nu-1}H_0^{\nu-1} + \cdots + c_1 H_0^1 + c_0 H_0^0\right)^{k_0}\right).$$

Let $B = c_{\nu-1}H_0^{\nu-1} + \cdots + c_1 H_0^1$. Since $H_0^0 = 1$, we get

$$\sum_{c_0 \in \mathbb{F}_q} \left(\left(c_{\nu-1}H_0^{\nu-1} + \cdots + c_1 H_0^1 + c_0 H_0^0 \right)^{k_0+l_0} \right.$$

$$\left. - \delta_0 \left(c_{\nu-1}H_0^{\nu-1} + \cdots + c_1 H_0^1 + c_0 H_0^0 \right)^{k_0} \right)$$

$$= \sum_{c_0 \in \mathbb{F}_q} (B + c_0)^{k_0+l_0} - \delta_0 \sum_{c_0 \in \mathbb{F}_q} (B + c_0)^{k_0} \stackrel{\text{def}}{=} \Sigma_1 - \delta_0 \Sigma_2.$$

We find that

$$\Sigma_2 = B^{k_0} + \sum_{j=0}^{k_0} \binom{k_0}{j} B^{k_0-j} \sum_{0 \neq c_0 \in \mathbb{F}_q} c_0^j = \begin{cases} 0, & \text{if } k_0 \neq q-1; \\ -1, & \text{if } k_0 = q-1, \end{cases}$$

since for $0 \leq j \leq k_0 \leq q-1$ we have

$$\sum_{0 \neq c_0 \in \mathbb{F}_q} c_0^j = \begin{cases} -1, & \text{if } j = 0, q-1; \\ 0, & \text{otherwise.} \end{cases}$$

Similarly,

$$\Sigma_1 = B^{k_0+l_0} + \sum_{j=0}^{k_0+l_0} \binom{k_0+l_0}{j} B^{k_0+l_0-j} \sum_{0 \neq c_0 \in \mathbb{F}_q} c_0^j$$

$$= \begin{cases} 0, & \text{if } 0 \leq k_0 + l_0 < q-1; \\ -\binom{k_0+l_0}{q-1} B^{k_0+l_0-q+1}, & \text{if } q-1 \leq k_0 + l_0 < 2q-2; \\ -\binom{k_0+l_0}{q-1} B^{q-1} - 1, & \text{if } k_0 + l_0 = 2q-2, \end{cases}$$

$$= \begin{cases} 0, & \text{if } 0 \leq k_0 + l_0 < q-1; \\ -1, & \text{if } k_0 + l_0 = q-1; \\ -1, & \text{if } 0 \leq k_0 + l_0 = 2q-2. \end{cases}$$

Therefore

$$\Sigma_1 - \delta_0 \Sigma_2 = \begin{cases} 0, & \text{if } k_0 + l_0 \neq q-1; \\ -1, & \text{if } k_0 + l_0 = q-1. \end{cases}$$

Now, if $k_0 + l_0 = q - 1$, then

$$\sum_{\deg m < \nu} G_k(m) g_l(m)$$

$$= - \sum_{c_1,\ldots,c_{\nu-1} \in \mathbb{F}_q} \prod_{i=1}^{\nu-1} \left(\left(c_{\nu-1} H_i^{\nu-1-i} + \cdots + c_{i+1} H_i^1 + c_i H_i^0 \right)^{k_i + l_i} \right.$$

$$\left. - \delta_i \left(c_{\nu-1} H_i^{\nu-1-i} + \cdots + c_{i+1} H_i^1 + c_i H_i^0 \right)^{k_i} \right).$$

Summing up on $c_1, \ldots, c_{\nu-1}$, we get the identity

$$\sum_{\deg m < \nu} G_k(m) g_l(m) = \begin{cases} (-1)^\nu, & \text{if } k_0 + l_0 = \ldots = k_{\nu-1} + l_{\nu-1} = q - 1; \\ 0, & \text{otherwise,} \end{cases}$$

which coincides with (1.45).

(iv) The proof of the identity (1.46) is similar to that of (1.45). ■

We use the orthogonality relation (1.45) to obtain an explicit formula for the coefficients of the expansion

$$f(t) = \sum_{i=0}^{\infty} a_i G_i(t), \quad t \in O, \tag{1.49}$$

of an arbitrary function $f \in C(O, K)$.

Theorem 1.15 *The coefficients $a_i \to 0$ of the expansion (1.49) are expressed as follows:*

$$a_i = (-1)^\nu \sum_{\deg m < \nu} g_{q^\nu - i - 1}(m) f(m), \tag{1.50}$$

for any integer ν, such that $q^\nu > i$.

Proof. By (1.49), the right-hand side of (1.50) equals

$$\sum_{i=0}^{\infty} a_i (-1)^\nu \sum_{\deg m < \nu} g_{q^\nu - i - 1}(m) G_i(m),$$

which is equal to a_i by virtue of (1.45). ■

Remark 1.2 Following Remark 1.1(4) of Sect. 1.2.1, we conclude from the formulas for general Carlitz polynomials and the expression (1.50) that the sequence $\{G_i\}_0^\infty$ forms an $\mathbb{A}$-basis of the $\mathbb{A}$-module $\text{Int}(\mathbb{A})$ of all integer-valued polynomials from $\mathbb{A}[t]$. The coefficients of the expansion of polynomials in the basis $\{G_i\}$ can be obtained by (1.50) in this situation too.

It is well known (see, for example, [77], p. 420) that
$$\binom{p-1}{n} \equiv (-1)^n \pmod{p}, \tag{1.51}$$
if $0 \leq n \leq p-1$. Using (1.47) we find from (1.51) that for any $\nu \in \mathbb{N}$
$$\binom{q^\nu - 1}{n} \equiv (-1)^n \pmod{p}, \tag{1.52}$$
if $0 \leq n \leq q^\nu - 1$. The congruence (1.52) leads to the following corollary to Proposition 1.14.

Corollary 1.16 *The polynomials $G_{q^\nu-1}$ and $g_{q^\nu-1}$ satisfy the following identities:*
$$G_{q^\nu-1}(t+s) = \sum_{j+l=q^\nu-1} (-1)^j G_j(t) G_l(s);$$
$$G_{q^\nu-1}(t-s) = \sum_{j+l=q^\nu-1} G_j(t) G_l(s);$$
$$g_{q^\nu-1}(t+s) = \sum_{j+l=q^\nu-1} (-1)^j G_j(t) g_l(s);$$
$$g_{q^\nu-1}(t-s) = \sum_{j+l=q^\nu-1} G_j(t) g_l(s).$$

The polynomial system $\{g_{q^\nu-1}\}$ has another interesting property.

Proposition 1.17 *The system of polynomials $g_{q^\nu-1}(t)$, $t \in O$, $\nu = 0, 1, 2, \ldots$, is orthonormal.*

Proof. By Corollary 1.16,
$$g_{q^\nu-1}(t) = \sum_{j=0}^{q^\nu-1} G_j(t) g_{q^\nu-1-j}(0) \tag{1.53}$$
where $g_0(0) = 1$. On the other hand, by the definition of g_j,
$$g_{q^\nu-1}(t) = \prod_{i=0}^{\nu-1} \left(f_i^{q-1}(t) - 1 \right). \tag{1.54}$$

It follows from (1.54) that $\|g_{q^\nu-1}\| \leq 1$ and $g_{q^\nu-1}(0) = (-1)^\nu$, so that $\|g_{q^\nu-1}\| = 1$ for all ν. In order to prove orthogonality, it suffices to show that for any $n \in \mathbb{N}$, $\lambda_1, \ldots, \lambda_n \in K$,

$$\left\| \sum_{k=0}^{n} \lambda_k g_{q^k-1} \right\| \geq |\lambda_n|$$

(see the proof of Proposition 50.4 in [98]). By (1.53),

$$\sum_{k=0}^{n} \lambda_k g_{q^k-1}(t) = \sum_{j=0}^{q^n-1} G_j(t) \sum_{\log_q(j+1) \leq k \leq n} \lambda_k g_{q^k-j-1}(0).$$

Since the sequence $\{G_j\}$ is orthonormal,

$$\left\| \sum_{k=0}^{n} \lambda_k g_{q^k-1} \right\| = \max_j \left| \sum_{\log_q(j+1) \leq k \leq n} \lambda_k g_{q^k-j-1}(0) \right| \geq |\lambda_n g_0(0)| = |\lambda_n|$$

(for $j = q^n - 1$ the sum consists of a single term), as desired. ∎

1.4.3. Extended hyperdifferentiations. It is easy to see that the results of Section 1.4.2 remain valid if we consider, instead of the general Carlitz polynomials, the general construction (1.41), provided

$$\varphi_j(x^k) = \begin{cases} 0, & \text{if } k < j, \\ 1, & \text{if } k = j. \end{cases}$$

In particular, this applies to the case where $\varphi_j = \mathcal{D}_j$. The sequence of functions obtained from the hyperdifferentiations $\mathcal{D}_j$ via the digit construction (1.41) ("extended hyperdifferentiations") satisfy all the assertions similar to those of Proposition 1.14, Theorem 1.15, Corollary 1.16 and Proposition 1.17. The role of the polynomials g_j is played in this case by the functions

$$d_i(t) = \prod_{j=0}^{n-1} d_{\alpha_j q^j}(t)$$

where, just as above,

$$i = \alpha_0 + \alpha_1 q + \cdots + \alpha_{n-1} q^{n-1}, \quad 0 \leq \alpha_j \leq q-1,$$

$$d_{\alpha_j q^j}(t) = \begin{cases} \mathcal{D}_j^{\alpha_j}(t), & \text{if } 0 \leq \alpha_j < q-1; \\ \mathcal{D}_j^{\alpha_j}(t) - 1, & \text{if } \alpha_j = q-1. \end{cases}$$

1.5 Finite places of a global function field

1.5.1. Main notions. While classical analysis usually deals with the fields $\mathbb{R}$ and $\mathbb{C}$, in algebraic number theory the main objects are global fields and objects defined over them, like quadratic forms, diophantine equations, various number-theoretic functions, etc. In characteristic zero, the global fields are the field $\mathbb{Q}$ of rational numbers and its finite algebraic extensions. In positive characteristic p, the global fields are finite separable extensions of the function field $\mathbb{F}_q(x)$ consisting of all rational functions of x with coefficients from $\mathbb{F}_q$. The local field K considered above is one of the possible completions of $\mathbb{F}_q(x)$.

It is clear that global fields are objects of a discrete nature. Since this book is devoted to analysis on local fields, the global fields remain essentially outside its scope. Nevertheless in this section we introduce some basic notions regarding completions of $\mathbb{F}_q(x)$, to be able to mention some global aspects of the objects we consider.

Let λ be an isomorphic imbedding of $\mathbb{F}_q(x)$ into a local field L. A couple (λ, L) is called *a completion* of $\mathbb{F}_q(x)$, if $\lambda(\mathbb{F}_q(x))$ is dense in L. Two completions (λ, L) and (λ', L') are called *equivalent*, if there exists an isomorphism ρ of the field L onto L' such that $\lambda' = \rho \circ \lambda$. An equivalence class of completions of the field $\mathbb{F}_q(x)$ is called *a place*.

There are the following explicit constructions of completions of $\mathbb{F}_q(x)$. Let $0 \neq t \in \mathbb{F}_q(x)$; then $t = \dfrac{\theta_1}{\theta_2}$, $\theta_1, \theta_2 \in \mathbb{F}_q[x]$. Set

$$|t|_\infty = q^{\deg \theta_1 - \deg \theta_2}, \quad |0|_\infty = 0.$$

It is easy to check that $|\cdot|_\infty$ is a non-Archimedean absolute value on $\mathbb{F}_q(x)$, it defines a metric on $\mathbb{F}_q(x)$, and the completion K_∞ of $\mathbb{F}_q(x)$ with respect to that metric is a local field. Thus, the absolute value $|\cdot|_\infty$ defines a place of $\mathbb{F}_q(x)$ often called *the infinite place*. Note that $|x|_\infty = q$, $|x^{-1}|_\infty = q^{-1}$, so that x^{-1} is a prime element of K_∞ (see [17, 38, 122]) for various notions regarding local fields).

Another construction of places of $\mathbb{F}_q(x)$ is as follows. Let $\pi \in \mathbb{F}_q[x]$ be a monic irreducible polynomial, $\deg \pi = \delta \geq 1$. If $t \in \mathbb{F}_q(x)$, write $t = \pi^n \dfrac{\tau}{\tau'}$ where $n \in \mathbb{Z}$, $\tau, \tau' \in \mathbb{F}_q[x]$, and π does not divide τ, τ'. In this case we also use the notations $n = \operatorname{ord}_\pi t$ and $t \equiv 0 \pmod{\pi^n}$. Set $|0|_\pi = 0$,

$$|t|_\pi = |\pi|_\pi^n, \quad |\pi|_\pi = q^{-\delta}.$$

We will denote by K_π the completion of $\mathbb{F}_q(x)$ with respect to the metric determined by this absolute value. Again, this is a non-Archimedean local field. For $\pi(x) \equiv x$, we obtain the field K considered in the preceding sections.

Denote by O_π the ring of integers, and by P_π the prime ideal in K_π, that is

$$O_\pi = \{t \in K_\pi : |t|_\pi \leq 1\}, \quad P_\pi = \{t \in K_\pi : |t|_\pi < 1\}.$$

Theorem 1.18 (i) *The local fields K_∞ and K_π give the complete list of different places of the global field $\mathbb{F}_q(x)$.*

(ii) *For each local field K_π, the cardinality of its residue field $\mathbb{F}_\pi = O_\pi/P_\pi$ equals q^δ, and a full system of representatives of the residue classes consists of all polynomials from $\mathbb{F}_q[x]$ of degrees $< \delta$.*

For the proof see [122] (note that our notation is slightly different from that in [122]).

The places of $\mathbb{F}_q(x)$ corresponding to the completions K_π are often called *finite*. Of course, all the fields K_π and K_∞ are isomorphic to K; to obtain the isomorphism, it suffices to take the prime element (respectively, π and x^{-1}) for the new variable. Thus, the global field structures do not bring anything new on the local field level. However they lead to quite different properties of objects defined in terms of a global field; in particular, the infinite place is usually very different from the finite ones, while properties of finite places are often similar to each other.

1.5.2. The Carlitz polynomials. It is clear from (1.11) or (1.14) that the Carlitz polynomials e_i and the normalized Carlitz polynomials f_i are defined globally, they belong to $\mathbb{F}_q[x][t]$ and $\mathbb{F}_q(x)[t]$, respectively. In this section we prove that the system $\{f_i\}$ and its extension by the digit principle form orthonormal bases in Banach spaces associated with all the finite places of $\mathbb{F}_q(x)$. First we need the following auxiliary result.

Lemma 1.19 *Let π be irreducible in $\mathbb{F}_q[x]$, $\deg \pi = \delta$. If $j < \delta n$, then*

$$f_j(\pi^n g) \equiv 0 \pmod{\pi}$$

for all $g \in \mathbb{F}_q[x]$.

Proof. We may suppose that $g \neq 0$, and we have to show that

$$\operatorname{ord}_\pi (e_j(\pi^n g)) > \operatorname{ord}_\pi(D_j).$$

For an integer $k \geq 0$, let $k \equiv R_k \pmod{\delta}$, where $0 \leq R_k \leq \delta - 1$. In particular, we write $j = \delta Q + R_j$.

By Lemma 2.13 from [76],

$$\operatorname{ord}_\pi([n]) = \begin{cases} 1, & \text{if } \delta \text{ divides } n; \\ 0, & \text{otherwise.} \end{cases} \tag{1.55}$$

From (1.55) and the definition (1.1) of D_k we get by an easy computation that

$$\operatorname{ord}_\pi(D_k) = \frac{q^k - q^{R_k}}{q^\delta - 1}. \tag{1.56}$$

Since $\deg(\pi^n g) > j$, we find from (1.56) and Lemma 1.1(iii) that

$$\operatorname{ord}_\pi (e_j(\pi^n g)) = n + \operatorname{ord}_\pi(g) + \sum_{k=0}^{j-1}(q-1)\operatorname{ord}_\pi(D_k)$$

$$= n + \operatorname{ord}_\pi(g) + \sum_{k=0}^{j-1}(q-1)\frac{q^k - q^{R_k}}{q^\delta - 1}$$

$$= n + \operatorname{ord}_\pi(g) + \frac{q^j - q^{R_j}}{q^\delta - 1} - Q$$

$$= n + \operatorname{ord}_\pi(g) + \operatorname{ord}_\pi(D_j) - Q > \operatorname{ord}_\pi(D_j),$$

since $n > \frac{j}{\delta} \geq Q$. ∎

Let $C_0(O_\pi, K_\pi)$ be the vector space over K_π of all continuous $\mathbb{F}_q$-linear functions $O_\pi \to K_\pi$, endowed with the supremum norm.

Theorem 1.20 *For any irreducible polynomial $\pi \in \mathbb{F}_q[x]$, $\delta = \deg \pi \geq 1$, the polynomials $f_j(t)$, viewed as elements of $C_0(O_\pi, K_\pi)$, form an orthonormal basis in $C_0(O_\pi, K_\pi)$.*

Proof. Let n be an arbitrary natural number. By Lemma 1.19, for $j < \delta n$ the reduction $\overline{f_j}$ is a well-defined mapping from $\mathbb{F}_q[x]/(\pi^n)$ to $\mathbb{F}_q[x]/(\pi) \cong \mathbb{F}_\pi$. Here (π^n) and (π) are the principal ideals in $\mathbb{F}_q[x]$ generated by π^n and π, respectively.

For $0 \leq j, k \leq \delta n - 1$, the $\delta n \times \delta n$ matrix $(f_j(x^k))$ is triangular, with all diagonal entries equal to 1. Since $\{1, x, \ldots, x^{\delta n - 1}\}$ is an $\mathbb{F}_q$-basis of

$O_\pi/(\pi^n) \cong \mathbb{F}_q[x]/(\pi^n)$, it follows that the δn reduced functions $\overline{f_j}$ form an $\mathbb{F}_\pi$-basis of $\mathrm{Hom}_{\mathbb{F}_q}(\mathbb{F}_\pi[[\pi]]/(\pi^n), \mathbb{F}_\pi)$. Using Proposition 1.5 we come to the conclusion that $\{f_j\}$ is an orthonormal basis in $C_0(O_\pi, K_\pi)$. ∎

Of course, for $\pi(x) = x$ the assertion of Theorem 1.20 coincides with the first assertion of Theorem 1.8. Thus, we have just presented a second proof of the latter.

An obvious task now is to construct an orthonormal basis of the Banach space $C(O_\pi, K_\pi)$ of all the continuous K_π-valued functions on O_π. The digit principle does not carry over in a straightforward way, because $C_0(O_\pi, K_\pi)$ consists, by definition, of all $\mathbb{F}_q$-linear functions, while the residue field $\mathbb{F}_\pi = O_\pi/K_\pi$ is in general bigger than $\mathbb{F}_q$. In spite of this discrepancy, the following result is valid.

Theorem 1.21 *The system of general Carlitz polynomials is an orthonormal basis of the space $C(O_\pi, K_\pi)$.*

Proof. As before, we denote $\delta = \deg \pi \geq 1$. Let $\overline{f_j} : O_\pi \to \mathbb{F}_\pi$ be reductions of the normalized Carlitz polynomials f_j. As in the proof of Theorem 1.13, denote

$$H_n = \bigcap_{j=0}^{\delta n - 1} \ker \overline{f_j}.$$

We already know that $\{\overline{f_0}, \ldots, \overline{f_{\delta n - 1}}\}$ is an $\mathbb{F}_\pi$-basis of the set of all $\mathbb{F}_q$-linear maps from $\mathbb{F}_q[x]/(\pi^n)$ to $\mathbb{F}_q[x]/(\pi)$. Therefore these functions separate the points of $\mathbb{F}_q[x]/(\pi^n)$. Now the argument from the proof of Theorem 1.13 shows that the set $\mathrm{Maps}(O_\pi/H_n, \mathbb{F}_\pi)$ is spanned over $\mathbb{F}_\pi$ by the monomials

$$\overline{f_0}^{\beta_0} \cdots \overline{f_{\delta n - 1}}^{\beta_{\delta n - 1}}, \quad 0 \leq \beta_j \leq q^\delta - 1.$$

Note that any $f_j^{q^k}$ is $\mathbb{F}_q$-linear, so in $\mathrm{Maps}(O_\pi/H_n, \mathbb{F}_\pi)$ we can write $\overline{f_j}^{q^k}$ as an $\mathbb{F}_\pi$-linear combination of $\{\overline{f_0}, \ldots, \overline{f_{\delta n - 1}}\}$. Therefore for all $n \geq 1$, $\mathrm{Maps}(O_\pi/H_n, \mathbb{F}_\pi)$ is spanned over $\mathbb{F}_\pi$ by the monomials

$$\overline{f_0}^{\gamma_0} \cdots \overline{f_{\delta n - 1}}^{\gamma_{\delta n - 1}}, \quad 0 \leq \gamma_j \leq q - 1,$$

so this set is an $\mathbb{F}_\pi$-basis. It remains to use Proposition 1.5. ∎

The explicit expressions for the coefficients of expansions in the Carlitz polynomials remain valid in the above global field framework (for finite places). However we have to stress that all the above results fail for the completion K_∞; the Carlitz polynomials do not even take integral elements to integral elements at this place. For the use of the Carlitz polynomials in an investigation of entire functions on K_∞ see [21].

1.5.3. Hyperdifferentiations. Let us consider, in the framework of Sections 1.5.1 and 1.5.2, the hyperdifferentiations $\mathcal{D}_k(t)$ investigated in Sections 1.3 and 1.4.3.

The Leibnitz rule (1.29) extends to the case of more than two factors, as

$$\mathcal{D}_k(t_1 \cdots t_m) = \sum_{\substack{j_1 + \cdots j_m = k \\ j_1, \ldots, j_m \geq 0}} \mathcal{D}_{j_1}(t_1) \cdots \mathcal{D}_{j_m}(t_m).$$

It follows that for any polynomials $t, s \in \mathbb{F}_q[x]$,

$$\mathcal{D}_k(t^n s) \equiv 0 \pmod{t^{n-k}}, \quad n \geq k. \tag{1.57}$$

If, as above, K_π is a completion of $\mathbb{F}_q(x)$ corresponding to a monic irreducible polynomial, then (1.57) implies the continuity of $\mathcal{D}_k(t)$ as a function on O_π with values in K_π. It can be proved [29] that the functions $\mathcal{D}_k$ are actually not only $\mathbb{F}_q$-linear but even $\mathbb{F}_\pi$-linear. The sequence $\{\mathcal{D}_k(t)\}_{k=0}^{\infty}$ is an orthonormal basis of the space of all $\mathbb{F}_\pi$-linear continuous functions $O_\pi \to K_\pi$. In order to construct an orthonormal basis of $C(O_\pi, K_\pi)$, one can use Theorem 1.13. The resulting basis is different from the basis of extended hyperdifferentiations considered in Sect. 1.4.3, because here the construction (similar to (1.41)) must use the q^δ-base representation of integers, not the q-base representation as before.

Just as for the Carlitz polynomials, the infinite place does not fit into the above picture. Though the functions $\mathcal{D}_k$ can be extended from $\mathbb{F}_q(x)$ onto K_∞, they are not orthonormal in this case. See [29] for further details regarding the global field treatment of hyperdifferentiations.

1.6 The Carlitz module

1.6.1. The Carlitz exponential and logarithm. The Carlitz exponential over K is defined by the power series

$$e_C(z) = \sum_{j=0}^{\infty} \frac{z^{q^j}}{D_j}. \tag{1.58}$$

It follows from (1.3) that the series (1.58) is convergent if $z \in \overline{K}_c$, $|z| < q^{-\frac{1}{q-1}}$.

Before writing down an explicit formula for the function inverse to e_C, we prove the following lemma. Let $\{a_n\}_0^\infty$, $\{b_n\}_0^\infty$ be sequences in $\overline{K}_c$.

Lemma 1.22 *Suppose that*

$$\sum_{i+j=k} a_i b_j^{q^i} = \begin{cases} 0, & \text{for } k > 0, \\ 1, & \text{for } k = 0. \end{cases} \quad (1.59)$$

Then

$$\sum_{i+j=k} b_j a_i^{q^j} = \begin{cases} 0, & \text{for } k > 0, \\ 1, & \text{for } k = 0. \end{cases} \quad (1.60)$$

Proof. For $k = 0$, both the equalities (1.59) and (1.60) mean that $a_0 b_0 = 1$. For $k = 1$, (1.59) means that $a_0 b_1 + a_1 b_0^q = 0$, whence $b_1 + a_1 b_0^{q+1} = 0$. Then

$$b_0 a_1 + b_1 a_0^q = a_0^q \left(b_1 + a_1 b_0^{q+1} \right) = 0,$$

and this is the equality required for $k = 1$ in (1.60).

Suppose that (1.60) is true for all $k < m$. Then, by (1.59),

$$\sum_{l=0}^{m-1} b_l a_{m-l}^{q^l} = -a_0^{q^m} \sum_{l=0}^{m-1} b_l \sum_{m-l=i+j} a_i^{q^l} b_j^{q^{l+j}}$$

$$= -a_0^{q^m} \sum_{\substack{m=l+i+j \\ j>0}} b_l a_i^{q^l} b_j^{q^{l+j}}$$

$$= -a_0^{q^m} \sum_{\substack{m=k+j \\ j>0}} \left\{ \sum_{k=l+j} b_l a_i^{q^l} \right\} b_j^{q^k}.$$

Since $j > 0$, $k < m$, the equality (1.60) is valid for the inner sum, and we find that

$$\sum_{l=0}^{m-1} b_l a_{m-l}^{q^l} = -a_0^{q^m} b_m,$$

so that (1.60) holds for $k = m$. This completes the proof. ■

The Carlitz logarithm $\log_C(z)$ is defined as

$$\log_C(z) = \sum_{n=0}^{\infty} (-1)^n \frac{z^{q^n}}{L_n}. \qquad (1.61)$$

Since $|L_n| = q^{-n}$, the series in (1.61) converges if $|z| < 1$. Below it will be considered for $z \in \overline{K}_c$, $|z| < q^{-1/(q-1)}$.

Theorem 1.23 *The $\mathbb{F}_q$-linear functions e_C and $\log_C$ are inverse to each other.*

Proof. We note first of all that e_C and $\log_C$ are mutually inverse as formal power series. Indeed, we have to prove that

$$\sum_{n=0}^{l} \frac{(-1)^{l-n}}{D_n L_{l-n}^{q^n}} = \begin{cases} 0, & \text{for } l > 0; \\ 1, & \text{for } l = 0, \end{cases} \qquad (1.62)$$

$$\sum_{n=0}^{l} \frac{(-1)^n}{L_n D_{l-n}^{q^n}} = \begin{cases} 0, & \text{for } l > 0; \\ 1, & \text{for } l = 0. \end{cases} \qquad (1.63)$$

The left-hand side of (1.62) equals $f_l(1)$ (see Proposition 1.7), so that the equalities (1.62) follow from (1.11) and (1.15). Now the equalities (1.63) are consequences of Lemma 1.22.

By (1.61) and the ultra-metric inequality,

$$|\log_C(z)| \le \sup_{n \ge 0} q^n |z|^{q^n}.$$

The function $\psi_z(s) = s|z|^s$ decreases for $s > -(\log|z|)^{-1}$; if $|z| < q^{-1/(q-1)}$, then ψ_z decreases for $s > \frac{q-1}{\log q}$. In particular, $\psi_z(q^n) \le \psi_z(q)$, $n \ge 1$. Hence, $|\log_C(z)| \le \max(|z|, q|z|^q)$, and we find that

$$|\log_C(z)| < q^{-\frac{1}{q-1}}, \quad \text{if } |z| < q^{-\frac{1}{q-1}}.$$

Therefore $e_C \circ \log_C$ is well-defined, and

$$e_C(\log_C(z)) = z, \quad \text{if } |z| < q^{-\frac{1}{q-1}}.$$

Similarly, $|e_C(z)| < q^{-\frac{1}{q-1}}$, if $|z| < q^{-\frac{1}{q-1}}$, and $\log_C(e_C(z)) = z$. ∎

It follows immediately from the definition of the Carlitz difference operators Δ, $\Delta^{(j)}$ (see (1.17), (1.18)) that

$$\Delta e_C(t) = e_C(t)^q, \quad \Delta^{(j)} e_C(t) = e_C(t)^{q^j}. \qquad (1.64)$$

If we introduce an $\mathbb{F}_q$-linear operator
$$d = \sqrt[q]{} \circ \Delta \qquad (1.65)$$
called *the Carlitz derivative*, then the first equality (1.64) takes the form
$$de_C = e_C. \qquad (1.66)$$

The equation (1.66) is the first example of a differential equation with Carlitz derivatives. See Chapter 3 for general results regarding such equations.

While the Carlitz exponential $e_C(t)$ is defined for $|t| < q^{-\frac{1}{q-1}}$, it is useful to introduce the function
$$w_z(t) = e_C(tz), \quad z \in \overline{K}_c, \ |z| < q^{-\frac{1}{q-1}},$$
defined for $t \in O$. Using Theorem 1.8 and the identities (1.64) we find an explicit expansion of the function w_z in the normalized Carlitz polynomials:
$$w_z(t) = \sum_{n=0}^{\infty} (e_C(z))^{q^n} f_n(t), \quad t \in O, \ |z| < q^{-\frac{1}{q-1}} \qquad (1.67)$$
(note that when we deal with expansions in the Carlitz polynomials, we have to consider only $t \in O \subset K$).

1.6.2. The Carlitz module. Consider *the Carlitz module function*
$$C_s(t) = \sum_{n=0}^{\infty} f_n(s) t^{q^n}, \quad s \in O, \ t \in \overline{K}_c, \ |t| < 1, \qquad (1.68)$$
$\mathbb{F}_q$-linear in each of its arguments s, t. If $s \in \mathbb{F}_q[x]$, then the series in (1.68) is actually a finite sum, from $n = 0$ to $n = \deg s$. In the general case, the convergence of the series follows from the fact that $\|f_n\| = 1$ for each $n = 0, 1, 2, \ldots$

It follows from (1.67) that
$$C_s(e_C(z)) = e_C(sz), \quad s \in O, \ |z| < q^{-\frac{1}{q-1}}. \qquad (1.69)$$

The identity (1.69) is the main functional equation for the Carlitz exponential. It can also be rewritten as a relation for the Carlitz logarithm:
$$\log_C(C_s(z)) = s \log_C(z).$$

If $s_1, s_2 \in O$, then by (1.69),
$$C_{s_1}(C_{s_2}(e_C(z))) = C_{s_1 s_2}(e_C(z)).$$

As we know from Theorem 1.23, the set of values of $e_C(z)$, $|z| < q^{-\frac{1}{q-1}}$, coincides with the disk $V = \left\{ z \in \overline{K}_c : |z| < q^{-\frac{1}{q-1}} \right\}$. Therefore

$$C_{s_1}(C_{s_2}(\zeta)) = C_{s_1 s_2}(\zeta) \tag{1.70}$$

for all $\zeta \in V$. In particular, if $s_1, s_2 \in \mathbb{F}_q[x]$, then C_{s_1} and C_{s_2} are $\mathbb{F}_q$-linear polynomials in ζ. In this case the identity (1.70) means that C_s defines a homomorphism of the ring $\mathbb{F}_q[x]$ into the composition ring $K\{\zeta\}$ of $\mathbb{F}_q$-linear polynomials with coefficients from K. The Carlitz module is the first nontrivial example of such a homomorphism. A general theory of such homomorphisms was initiated by Drinfeld [33], so that they are now called (under some natural assumptions) *the Drinfeld modules*. Note that analogs of the Carlitz exponential exist in very general situations.

The theory of Drinfeld modules is the most essential part of function field arithmetic; see [45, 111] for detailed expositions in a spirit quite different from the above approach. In the literature all the objects are considered in the global field setting, usually over K_∞. $C_s(z)$ is introduced only in the polynomial case, $s \in \mathbb{F}_q[x]$; on the other hand, instead of the ring $\mathbb{F}_q[x]$, much more general rings can be considered. Over K_∞, the Carlitz exponential is an entire function; a detailed study of its properties including a description of its periods (from an extension of K_∞) and special values, as well as their number-theoretic applications, is given in [45]. We do not try to repeat that material in this book, due to its more analytic and elementary character.

1.7 Canonical commutation relations

1.7.1. The background. In the quantum mechanics of harmonic oscillator (see e.g. [81, 106]) a creation operator a^+ transforms a stationary state into a stationary state of the next (higher) energy level; an annihilation operator a^- acts in the opposite way. The operators satisfy the canonical commutation relation (CCR) $a^- a^+ - a^+ a^- = I$. In quantum field theory these properties are used to obtain operators which change the number of particles.

The most widely used representation of the CCR is the Schrödinger representation, in which $a^\pm$ are operators on the Hilbert space $L_2(\mathbb{R})$. If the mass, frequency and the Planck constant are all taken equal to 1, then

$$a^\pm = \frac{1}{\sqrt{2}} \left(t \mp \frac{d}{dt} \right)$$

where we identify the variable t and the operator of multiplication by t.

Let $\{h_n\}_0^\infty$ be the system of Hermite functions, an orthonormal basis of $L_2(\mathbb{R})$. The operators $a^\pm$ act on this basis as follows:

$$a^+ h_n = \sqrt{n+1}\, h_{n+1}, \quad a^- h_n = \sqrt{n}\, h_{n-1}, \quad n = 0, 1, \ldots, \tag{1.71}$$

where $h_{-1} = 0$. For "the number operator" $a^+ a^-$ we have

$$a^+ a^- h_n = n h_n, \quad n = 0, 1, \ldots \tag{1.72}$$

The above relations hold also for the Bargmann–Fock representation of the CCR. Here the operators $\tilde{a}^\pm$ (satisfying the same relation $\tilde{a}^- \tilde{a}^+ - \tilde{a}^+ \tilde{a}^- = I$) act on the Hilbert space of entire functions

$$u(z) = \sum_{n=0}^\infty c_n \frac{z^n}{\sqrt{n!}}, \quad z \in \mathbb{C}, \ \sum_{n=0}^\infty |c_n|^2 < \infty,$$

with the inner product

$$(u_1, u_2) = \frac{1}{\pi} \int_\mathbb{C} u_1(z) \overline{u_2(z)} e^{-|z|^2}\, dz.$$

Here $(\tilde{a}^+ u)(z) = z u(z)$, $(\tilde{a}^- u)(z) = u'(z)$. Instead of the Hermite functions, the orthonormal basis with the above properties is $\left\{\dfrac{z^n}{\sqrt{n!}}\right\}_{n=0}^\infty$.

An important object related to the CCR is the system of *coherent states*, generalized eigenfunctions (not necessarily belonging to the Hilbert space) of the annihilation operator. In the Bargmann–Fock representation these are the functions $z \mapsto e^{\lambda z}$, $\lambda \in \mathbb{C}$.

Analogs of the above constructions are known also in p-adic analysis. An analog of the Schrödinger representation [60] is as follows. The operators $a^\pm$ act on the space $C(\mathbb{Z}_p, \mathbb{Q}_p)$ of p-adic-valued continuous functions on the ring $\mathbb{Z}_p$ of p-adic integers,

$$(a^+ u)(t) = t u(t-1), \quad (a^- u)(t) = u(t+1) - u(t), \quad t \in \mathbb{Z}_p. \tag{1.73}$$

Note that it is a purely non-Archimedean phenomenon that a compact set $\mathbb{Z}_p$ is preserved under the unit shift of the argument in (1.73). Instead of (1.71)–(1.72) we have in this case

$$a^- P_n = P_{n-1}, \ n \geq 1; \quad a^- P_0 = 0,$$

$$a^+ P_n = (n+1) P_{n+1}, \quad n \geq 0,$$

so that $(a^+a^-)P_n = nP_n$ and, as before, $a^-a^+ - a^+a^- = I$. Here $\{P_n\}$ is the Mahler basis of $C(\mathbb{Z}_p, \mathbb{Q}_p)$ (see [98]), that is

$$P_n(t) = \frac{t(t-1)\cdots(t-n+1)}{n!}, \ n \geq 1; \quad P_0(t) \equiv 1.$$

The analogs of coherent states are the functions $f_\lambda(t) = (1+\lambda)^t$, where $|\lambda|_p < 1$. Various p-adic analogs of the Bargmann–Fock representation are given in [2, 57, 60]. It is interesting that, in contrast to the classical case, all the above p-adic operators $a^\pm$ are bounded. The problem of existence of linear bounded representations of CCR over valued fields of arbitrary rank is investigated in [56].

Below we construct a characteristic p representation of CCR. Just as in the above characteristic zero cases, the basic objects happen to be related to the main special functions of the corresponding branch of analysis – this time, to the Carlitz polynomials and the Carlitz exponential.

1.7.2. The representations. Denote $\tau u = u^q$. As before, we write $d = \sqrt[q]{} \circ \Delta$. All the operators below are considered on the space $C_0(O, \overline{K}_c)$ of $\mathbb{F}_q$-linear continuous functions on O with values in $\overline{K}_c$.

Theorem 1.24 *Let $a^+ = \tau - I$, $a^- = d$. Then*

$$a^-a^+ - a^+a^- = [1]^{1/q}I. \tag{1.74}$$

The operator a^+a^- possesses an orthonormal eigenbasis consisting of the normalized Carlitz polynomials:

$$(a^+a^-)f_i = [i]f_i, \quad i = 0, 1, 2, \ldots; \tag{1.75}$$

a^+ and a^- act upon the basis as follows:

$$a^+ f_{i-1} = [i]f_i, \quad a^- f_i = f_{i-1}, \ i \geq 1; \quad a^- f_0 = 0. \tag{1.76}$$

The equation

$$a^- u = \lambda u \tag{1.77}$$

has solutions ("coherent states") for any $\lambda \in \overline{K}_c$; if $\lambda \neq 0$, then each solution can be written as

$$u(t) = \lambda^{-\frac{q}{q-1}} \sum_{n=0}^{\infty} c^{q^n} f_n(t), \quad c \in \overline{K}_c, \ |c| < 1, \tag{1.78}$$

for some value of the $(q-1)$-th root, and conversely, $a^-u = \lambda u$ for the function (1.78). If in (1.78) $|c| < q^{-\frac{1}{q-1}}$, then

$$u(t) = \lambda^{-\frac{q}{q-1}} e_C(tz), \quad z = \log_C(c). \tag{1.79}$$

In particular, if $q \neq 2$, then every function (1.78) with $c \in K$ takes the form (1.79).

Proof. The identities (1.74)–(1.76) are checked by a direct computation, with the use of the properties (1.12) and (1.21) of the Carlitz polynomials.

Let u be a solution of (1.77). We can write

$$u(t) = \sum_{n=0}^{\infty} c_n f_n(t), \quad t \in O, \ c_n \to 0.$$

Applying the operator d, we get that

$$\lambda u = \sum_{n=1}^{\infty} c_n^{1/q} f_{n-1},$$

and, by the uniqueness of the expansion, that

$$c_{n+1}^{1/q} = \lambda c_n, \quad n = 0, 1, \ldots,$$

whence

$$c_n = \lambda^{q^n + q^{n-1} + \cdots + q} c_0^{q^n} = \mu^{-1}(c_0 \mu)^{q^n}, \quad n = 1, 2, \ldots,$$

where $\mu = \lambda^{\frac{q}{q-1}}$. Since $c_n \to 0$, we have $|c_0 \mu| < 1$, and we obtain the representation (1.78) with $c = c_0 \mu$. The representation (1.79) follows from (1.67). If $q \neq 2$, $c \in K$, $|c| < 1$, then $|c| \leq q^{-1} < q^{-1/(q-1)}$, so that in this case (1.78) is equivalent to (1.79). ■

Note that the operators $a^\pm$ are obviously continuous on $C_0(O, \overline{K}_c)$ but, in contrast to both the classical and p-adic cases, are not linear, only $\mathbb{F}_q$-linear.

In order to construct an analog of the Bargmann–Fock representation (a representation by operators on a space of holomorphic functions), consider a Banach space H over the field $\overline{K}_c$ consisting of $\mathbb{F}_q$-linear power series

$$u(t) = \sum_{n=0}^{\infty} a_n \frac{t^{q^n}}{D_n}, \quad a_n \in \overline{K}_c, \ |a_n| \to 0. \tag{1.80}$$

The norm in H is given by

$$\|u\|_H = \sup_n |a_n|.$$

It follows from (1.3) that a seies (1.80) defines a holomorphic function for $|t| < q^{-1/(q-1)}$. It is obvious that the sequence of functions $\tilde{f}_n(t) = \dfrac{t^{q^n}}{D_n}$, $n = 0, 1, 2, \ldots$, is an orthonormal basis of H.

The desired representation is given by the following operators on the space H:

$$\tilde{a}^+ = \tau, \quad \tilde{a}^- = d.$$

Note that the form of the operators $\tilde{a}^\pm$ is only slightly different from that of $a^\pm$, but they act on a different Banach space.

By a straightforward computation based on the identities

$$\Delta\left(t^{q^n}\right) = [n]t^{q^n}, \quad D_{n+1} = [n+1]D_n^q,$$

we show that the relations (1.74)–(1.76) hold for the operators $\tilde{a}^\pm$, with $\tilde{f}_n$ substituted for f_n.

1.8 Comments

The factorial-like sequences (1.1), (1.2), and (1.4) were introduced, among other basic constructions of the analysis in positive characteristic, by Carlitz [22, 23]. Studying their properties (Lemma 1.1) Carlitz used a different technique (the Moore determinants); see also [45]. Our exposition follows, in a little more detailed form, the proof of Proposition 3.1.6 in [45] (see also Theorem 3.20 in [76]). Proposition 1.2 is taken from [45].

Composition rings of locally holomorphic $\mathbb{F}_q$-linear functions and the corresponding skew fields of "meromorphic" functions (Propositions 1.3, 1.4) were studied by the author [66]. Note that a detailed investigation of bi-infinite series $\sum\limits_{k=-\infty}^{\infty} a_k t^{q^k}$ convergent on the whole of $\overline{K}_c$ was carried out by Poonen [85]. The results regarding orthonormal bases (Propositions 1.5, 1.6) are taken from [3, 101].

Properties (1.12)–(1.15) of the additive Carlitz polynomials were proved by Carlitz in the seminal paper [22] where the polynomials were introduced for the first time. Our proofs follow [45]. The identity (1.16) is due to Wagner [119]. In fact, Carlitz considered only expansions of polynomials, while the expansions of continuous functions in the system $\{f_i\}$ (Theorem 1.8) were first studied by Wagner [118, 119]. There are several different proofs of this result [118, 119, 43, 61, 29, 32]; we followed [119]. Theorem 1.10 was proved in [64].

Hyperdifferentiations were introduced by Hasse [48] and studied in a more general context in [108, 49, 99]; for various generalizations see [112] and references therein. An interpretation of hyperdifferentiations as functions on K was first given by Voloch [117] who proved Proposition 1.11 (i), (iii). Other results of Section 1.3 were obtained by Jeong [53, 55] and Snyder [105]. The connections (1.36), (1.39) of hyperdifferentiations with powers of the operator Δ were used by Jeong [54] to solve some difference equations containing Δ.

The digit construction (1.41) was first proposed by Carlitz [23] for the case of the Carlitz polynomials. In [23], Carlitz proved all the results of Section 1.4.2 (see also [43]) except Proposition 1.17 proved in [62]. Our proof of the crucial orthogonality relation (1.45) is taken from the dissertation by Yang [123]. It was noticed by Jeong [53] that all the constructions are applicable also to the hyperdifferentiations. The general digit principle (Theorem 1.13) was proved by Conrad [29] who also found its analog for the case of characteristic 0.

Apart from the above explicit constructions of bases for the positive characteristic case, there are also various general results, applicable both for the cases of our field K, the field $\mathbb{Q}_p$ of p-adic numbers, and its finite extensions. See [3, 7, 15, 19, 20, 29, 104, 107, 118]. Further constructions of orthonormal bases in $C_0(O, K)$ are provided by an appropriate version of the umbral calculus; see Chapter 2 below.

The global field interpretation of the Carlitz polynomials was initiated by Wagner [118]; see also [43, 29]. Our proofs of Theorems 1.20 and 1.21 are taken from [29] where the case of hyperdifferentiations is also considered in detail.

Note that the Carlitz expansions at finite places have been used by Goss [44] to study, in the spirit of the Iwasawa isomorphism, O_π-valued measures on O_π. See Section 2.6.

The notions of the Carlitz exponential and logarithm, as well as the Carlitz module identity (1.69) (for $s \in \mathbb{F}_q[x]$) were introduced by Carlitz [22]; his definitions had slightly different normalizations. We follow [43, 45].

The function field representations of the canonical commutation relations were constructed by the author [61, 62]. In this book we do not touch on non-Archimedean models in physics; see [57, 115].

2
Calculus

This chapter is devoted to function field counterparts of smoothness and analyticity, and their interplays with the rate of decay of the coefficients of the Fourier–Carlitz series. This is connected with the Carlitz difference operators and the Carlitz derivatives (see Chapter 1). The latter notion, via the appropriate notion of an antiderivative, leads to the Volkenborn-type integration theory for $\mathbb{F}_q$-linear functions. There is another extension that is a kind of fractional derivative. Wider classes of operators on $\mathbb{F}_q$-linear functions are introduced in the $\mathbb{F}_q$-linear umbral calculus and applied to construction of new orthonormal polynomial systems.

2.1 $\mathbb{F}_q$-Linear calculus

2.1.1. Taylor coefficients of $\mathbb{F}_q$-linear holomorphic functions. Let

$$u(t) = \sum_{n=0}^{\infty} c_n t^{q^n}, \quad c_n \in \overline{K}_c, \qquad (2.1)$$

and the series (2.1) have a nonzero radius of convergence.

The classical formula for the coefficients of a power series cannot be used to reconstruct the coefficients of (2.1) – in higher order terms, both the numerators, the higher derivatives, and the denominators, the factorials, vanish. The next result shows that the Carlitz difference operators (1.18) and the Carlitz factorials (1.1) are the appropriate replacements.

Theorem 2.1 *The coefficients of the series (2.1) can be found as follows:*

$$c_n = \frac{1}{D_n} \lim_{t \to 0} \frac{\Delta^{(n)} u(t)}{t^{q^n}}, \quad n = 0, 1, 2, \ldots \qquad (2.2)$$

Proof. It will be convenient to rewrite the series (2.1) as

$$u(t) = \sum_{n=0}^{\infty} a_n \frac{t^{q^n}}{D_n},$$

$a_n = c_n D_n$. We may also assume that $a_n \to 0$; the general case is reduced to this one by substituting λt for t with an appropriate λ (transforming the disk of convergence into the unit disk). In the notation of Section 1.7.2, now u belongs to the Banach space H.

If $\tilde{f}_n(t) = \dfrac{t^{q^n}}{D_n}$, then by the identities of the Bargmann–Fock-type representation of Section 1.7.2,

$$d\left(\lambda \tilde{f}_n\right) = \lambda^{1/q} \tilde{f}_{n-1}, \quad n \geq 1, \lambda \in \overline{K}_c; \quad d\tilde{f}_0 = 0.$$

It follows that

$$d^n u(t) = \sum_{k=n}^{\infty} a_k^{1/q^n} \frac{t^{q^{k-n}}}{D_{k-n}},$$

so that

$$a_k^{1/q^n} = \lim_{t \to 0} \frac{d^n u(t)}{t}. \tag{2.3}$$

On the other hand, from the identities (1.76) and (1.23) for the normalized Carlitz polynomials we find that

$$\tau^n d^n = \Delta^{(n)}. \tag{2.4}$$

Applying τ^n to both sides of (2.3) and using (2.4) we come to (2.2). ∎

2.1.2. Smooth functions. Let $u \in C_0(O, \overline{K}_c)$,

$$\mathfrak{D}^k u(t) = t^{-q^k} \Delta^{(k)} u(t), \quad t \in O \setminus \{0\}.$$

We will say that $u \in C_0^{k+1}(O, \overline{K}_c)$ if $\mathfrak{D}^k u$ can be extended to a continuous function on O.

$C_0^{k+1}(O, \overline{K}_c)$ can be considered as a Banach space over $\overline{K}_c$, with the norm

$$\sup_{t \in O} \left(|u(t)| + |\mathfrak{D}^k u(t)|\right).$$

Note that $C_0^1(O, \overline{K}_c)$ coincides with the set of all differentiable $\mathbb{F}_q$-linear functions $O \to \overline{K}_c$.

In this section we will obtain a characterization of functions from

$C_0^{k+1}(O, \overline{K}_c)$ in terms of coefficients of the expansion $u = \sum_{n=0}^{\infty} c_n f_n$. We begin with several auxiliary results. In particular, the next lemma, a necessary condition of the differentiability of a $\mathbb{F}_q$-linear function, is a special case of the main result of this section, which will be proved later.

Lemma 2.2 *If $u \in C_0^1(O, \overline{K}_c)$, then*

$$|c_n| q^n \longrightarrow 0, \quad n \to \infty. \tag{2.5}$$

Proof. First we prove that the sequence $\{|c_n| q^n\}$ is bounded. Assuming the opposite, we find a strictly increasing sequence $\{n_r\}$ such that $\lim_{r \to \infty} |c_{n_r}| q^{n_r} = \infty$. Choosing, if necessary, an appropriate subsequence, we may assume, for $0 \le n < n_r$, that

$$|c_n| q^n < |c_{n_r}| q^{n_r}. \tag{2.6}$$

Suppose that $u'(0) = \lambda$. It follows from the identity (1.54) that

$$t^{-1} f_n(t) = \frac{g_{q^n - 1}(t)}{L_n}, \quad n = 0, 1, 2, \ldots \tag{2.7}$$

(the polynomials at both sides of (2.7) have the same roots $t = x^l$, $l < n$, the same degrees and the same leading coefficients). If $\varphi(t) = t^{-1} u(t)$, then by (2.7),

$$\lim_{r \to \infty} \varphi(x^{n_r}) = \lim_{r \to \infty} \sum_{n=0}^{n_r} \frac{c_n}{L_n} g_{q^n - 1}(x^{n_r}) = \lambda. \tag{2.8}$$

On the other hand, by (2.7) and the representation (1.14) of the Carlitz polynomials, we find that

$$g_{q^n - 1}(x^n) = \frac{L_n}{D_n} \sum_{j=0}^{n} (-1)^{n-j} \begin{bmatrix} n \\ j \end{bmatrix} x^{(q^j - 1)n}$$

$$= (-1)^n + \sum_{j=1}^{n} (-1)^{n-j} \frac{L_n}{D_j L_{n-j}^{q^j}} x^{(q^j - 1)n}$$

where

$$\left| \frac{L_n}{D_j L_{n-j}^{q^j}} \right| = q^{-m_n},$$

$$m_n = n + (q^j - 1)n - \frac{q^j - 1}{q - 1} - q^j(n - j) = jq^j - (1 + q + \cdots + q^{j-1}) > 0,$$

so that $g_{q^n-1}(x^n) \equiv (-1)^n \pmod{x}$. Therefore

$$\lim_{r \to \infty} \left| \sum_{n=0}^{n_r} \frac{c_n}{L_n} g_{q^n-1}(x^{n_r}) \right|$$

$$= \lim_{r \to \infty} \left| \sum_{n=0}^{n_r-1} \frac{c_n}{L_n} g_{q^n-1}(x^{n_r}) + \frac{c_{n_r}}{L_{n_r}} ((-1)^{n_r} + B_r) \right|$$

where $|g_{q^n-1}(x^{n_r})| \leq 1$, $\left|\frac{c_n}{L_n}\right| = |c_n|q^n$, $|B_r| \leq q^{-1}$, $\left|\frac{c_{n_r}}{L_{n_r}}\right| = |c_{n_r}|q^{n_r} > |c_n|q^n$ (due to (2.6)), so that

$$\lim_{r \to \infty} \left| \sum_{n=0}^{n_r} \frac{c_n}{L_n} g_{q^n-1}(x^{n_r}) \right| = \lim_{r \to \infty} |c_{n_r}| q^{n_r} = \infty,$$

which contradicts (2.8).

Thus, the sequence $\{|c_n|q^n\}$ is bounded. Suppose that it does not tend to 0. Since we may multiply the coefficients c_n by a fixed power of x, and we may also change finitely many coefficients in an arbitrary way, we may assume that

$$|c_n|q^n \leq 1, \quad \limsup_{n \to \infty} |c_n|q^n = 1. \tag{2.9}$$

We have

$$\lim_{r \to \infty} \sum_{n=0}^{r} \frac{c_n}{L_n} g_{q^n-1}(x^r) = \lambda,$$

so that, given $\varepsilon > 0$, there exists $r_0 \in \mathbb{N}$, such that

$$\left| \sum_{n=0}^{r} \frac{c_n}{L_n} g_{q^n-1}(x^r) - \lambda \right| < \varepsilon, \quad r \geq r_0.$$

The above argument shows that

$$g_{q^n-1}(x^r) \equiv (-1)^n \pmod{x},$$

if $0 \leq n \leq r$. Therefore, assuming that $\varepsilon < 1$, we get

$$\sum_{n=0}^{r_0+n} (-1)^n \frac{c_n}{L_n} = \lambda + \sigma(n)$$

where $|\sigma(n)| < 1$ for all $n \geq 0$. Hence for any $n \geq 1$,

$$\left|(-1)^{r_0+n}\frac{C_{r_0+n}}{L_{r_0+n}}\right| = |\sigma(n) - \sigma(n-1)| < 1,$$

in contradiction to (2.9). ∎

The next result is an extension of Proposition 1.17 regarding the orthonormality of the sequence $\{g_{q^n-1}\}_{n=0}^{\infty}$.

Lemma 2.3 *If a function $v: O \setminus \{0\} \to \overline{K}_c$ is continuous and bounded, and v admits a pointwise convergent expansion*

$$v(t) = \sum_{n=0}^{\infty} v_n g_{q^n-1}(t), \quad t \in O \setminus \{0\}, \tag{2.10}$$

$v_n \in \overline{K}_c$, *then*

$$\sup_{t \in O \setminus \{0\}} |v(t)| = \sup_{0 \leq n \leq \infty} |v_n|.$$

Proof. It is clear that

$$|v(t)| \leq \sup_n |v_n|, \quad t \neq 0.$$

In order to prove the inverse inequality, we use the identity (1.46) (with $k=0$, $l=q^\nu-1$), which shows that

$$\sum_{\substack{\deg t = m \\ t \text{ monic}}} g_{q^n-1}(t) = \begin{cases} 0, & \text{if } n < m, \\ (-1)^m, & \text{if } n = m. \end{cases}$$

If $n > m$, then $g_{q^n-1}(t) = 0$ for $\deg t = m$, due to the identity (2.7) and the corresponding property of the Carlitz polynomials.

Now the summation in both sides of (2.10) yields

$$\sum_{\substack{\deg t = m \\ t \text{ monic}}} v(t) = (-1)^m v_m,$$

so that $|v_m| \leq \sup_{t \in O \setminus \{0\}} |v(t)|$. ∎

Lemma 2.4 *Let a function $w \in C(O, \overline{K}_c)$ be such that the function $\gamma(t) = tw(t)$ is $\mathbb{F}_q$-linear. Then*

$$w(t) = \sum_{n=0}^{\infty} w_n g_{q^n-1}(t), \quad w_n \in \overline{K}_c, w_n \to 0, \qquad (2.11)$$

and the series (2.11) is uniformly convergent on O.

Proof. Consider the expansion

$$\gamma(t) = \sum \gamma_n f_n(t), \quad t \in O, \qquad (2.12)$$

where $\gamma_n \in \overline{K}_c$, $\gamma_n \to 0$. The fact that the function $t^{-1}\gamma(t)$ is continuous at $t = 0$ means that $\gamma(t)$ is differentiable at $t = 0$. By Lemma 2.2, $|\gamma_n| q^n \to 0$ for $n \to \infty$.

Dividing both sides of (2.12) by t we obtain the expansion (2.11) with $w_n = L_n^{-1} \gamma_n$, so that $w_n \to 0$. ■

Now we are in a position to prove the characterization result.

Theorem 2.5 *The function $u = \sum_{n=0}^{\infty} c_n f_n \in C_0(O, \overline{K}_c)$ belongs to $C_0^{k+1}(O, \overline{K}_c)$ if and only if*

$$q^{nq^k} |c_n| \to 0 \quad \text{for } n \to \infty. \qquad (2.13)$$

In this case

$$\sup_{t \in O} |\mathfrak{D}^k u(t)| = \sup_{n \geq k} q^{(n-k)q^k} |c_n|. \qquad (2.14)$$

Proof. We use the identity

$$d^k f_n = \begin{cases} f_{n-k}, & \text{if } n \geq k, \\ 0, & \text{if } n < k. \end{cases}$$

Since $d^k = \sqrt[q^k]{} \circ \Delta^{(k)}$, we find that

$$\mathfrak{D}^k u(t) = t^{-q^k} \sum_{n=k}^{\infty} c_n f_{n-k}^{q^k}(t), \quad t \neq 0, \qquad (2.15)$$

which implies the identity

$$\left(\mathfrak{D}^k u(t)\right)^{q^{-k}} = \sum_{n=k}^{\infty} c_n^{q^{-k}} \frac{g_{q^{n-k}-1}(t)}{L_{n-k}}, \quad t \neq 0. \qquad (2.16)$$

Now, if (2.13) is satisfied, then we find from Proposition 1.17 that the right-hand side of (2.16) is a continuous function on O, which means that $u \in C_0^{k+1}(O, \overline{K}_c)$. The equality (2.14) follows from Lemma 2.3.

Conversely, suppose that $\mathfrak{D}^k u$ is continuous on O. Let $w(t) = \left(\mathfrak{D}^k u(t)\right)^{q^{-k}}$. By (2.15), the function $\gamma(t) = tw(t)$ has the form

$$\gamma(t) = \sum_{n=k}^{\infty} c_n^{q^{-k}} f_{n-k}(t), \quad t \neq 0. \tag{2.17}$$

Since w is continuous on O, we see that $\gamma(0) = 0$. On the other hand, $f_n(0) = 0$ for all n, so that the equality (2.17) holds for all $t \in O$, and the function γ is $\mathbb{F}_q$-linear (however, we cannot claim the uniform convergence of the series in (2.17)).

Now Lemma 2.4 implies the representation (2.11) with $w_n \to 0$. It follows from (2.11) and (2.17) that

$$\sum_{n=0}^{\infty} \left(w_n - c_{n+k}^{q^{-k}} L_n^{-1}\right) g_{q^n-1}(t) = 0$$

for all $t \neq 0$. By Lemma 2.3, this means that

$$w_n = c_{n+k}^{q^{-k}} L_n^{-1}$$

for $n \geq 0$, which implies (2.13). ∎

2.1.3. Indefinite sum. Viewing the operator d as a kind of derivative, it is natural to introduce an appropriate antiderivative. Following the terminology used in the analysis over $\mathbf{Z}_p$ (see [98]) we call it the indefinite sum.

Consider in $C_0(O, \overline{K}_c)$ the equation

$$du = f, \quad f \in C_0(O, \overline{K}_c). \tag{2.18}$$

Suppose that $f = \sum\limits_{k=0}^{\infty} \varphi_k f_k$, $\varphi_k \in \overline{K}_c$. Looking for $u = \sum\limits_{k=0}^{\infty} c_k f_k$ and using the fact that

$$du = \sum_{k=1}^{\infty} c_k^{1/q} f_{k-1} = \sum_{l=0}^{\infty} c_{l+1}^{1/q} f_l$$

we find that $c_{l+1} = \varphi_l^q$, $l = 0, 1, 2, \ldots$ Therefore u is determined by (2.18) uniquely up to the term $c_0 f_0(t) = c_0 t$, $c_0 = u(1)$.

Fixing $u(1) = 0$ we obtain an $\mathbb{F}_q$-linear bounded operator S on $C_0(O, \overline{K}_c)$, the operator of indefinite sum: $Sf = u$. It follows from Theorem 2.5 that

S is also a bounded operator on each space $C_0^k(O, \overline{K}_c)$, $k = 1, 2, \ldots$ Note that $Sf_k = f_{k+1}$, $k = 0, 1, \ldots$, so that S is not compact.

Another possible procedure to find Sf is an interpolation. If $u = Sf$ then
$$u(xt) - xu(t) = f^q(t), \quad u(1) = 0.$$

Setting successively $t = 1, x, x^2, \ldots$, we get
$$u(x) = f^q(1),$$
$$u(x^2) = f^q(x) + xf^q(1),$$
$$u(x^3) = f^q(x^2) + xf^q(x) + x^2 f^q(1),$$
$$\cdots\cdots\cdots\cdots\cdots\cdots\cdots\cdots\cdots\cdots\cdots\cdots\cdots$$
$$u(x^n) = f^q(x^{n-1}) + xf^q(x^{n-2}) + \ldots + x^{n-1} f^q(1).$$

Since u is assumed $\mathbb{F}_q$-linear, this determines $u(t)$ for all $t \in \mathbb{F}_q[x]$. Extending by continuity, we find $u(t)$ for any $t \in O$: if
$$t = \sum_{n=0}^{\infty} \zeta_n x^n, \quad \zeta_n \in \mathbb{F}_q,$$
then
$$u(t) = \sum_{n=0}^{\infty} \zeta_n u(x^n) = \sum_{n=1}^{\infty} \zeta_n \sum_{j=1}^{n} x^{j-1} f^q(x^{n-j}) = \sum_{j=1}^{\infty} \sum_{n=j}^{\infty} \zeta_n x^{j-1} f^q(x^{n-j}),$$
so that
$$u(t) = \sum_{n=0}^{\infty} x^n f^q \left(\sum_{m=0}^{\infty} \zeta_{m+n+1} x^m \right).$$

2.1.4. The Volkenborn-type integral. The Volkenborn integral of a function on $\mathbf{Z}_p$ was introduced in [116] (see also [98]), in order to obtain relations between some objects of p-adic analysis resembling classical integration formulas of real analysis. Here we extend this approach to the function field situation. Our definition is based essentially on the Carlitz difference operator Δ, which again shows its close connection with basic structures of analysis over the field K.

The integral of a function $f \in C_0^1(O, \overline{K}_c)$ is defined as
$$\int_O f(t)\, dt \stackrel{\text{def}}{=} \lim_{n \to \infty} \frac{Sf(x^n)}{x^n} = (Sf)'(0).$$

It is clear that the integral is an $\mathbb{F}_q$-linear continuous functional on $C_0^1(O, \overline{K}_c)$,
$$\int_O cf(t)\,dt = c^q \int_O f(t)\,dt, \quad c \in \overline{K}_c.$$

Since a power series $\sum_{n=0}^\infty a_n t^{q^n}$ with $a_n \to 0$ converges in $C_0^1(O, \overline{K}_c)$, it can be integrated termwise:
$$\int_O \sum_{n=0}^\infty a_n t^{q^n}\,dt = \sum_{n=0}^\infty a_n^q \int_O t^{q^n}\,dt.$$

The integral possesses the following "invariance" property (related, in contrast to the case of $\mathbf{Z}_p$, to the multiplicative structure):
$$\int_O f(xt)\,dt = x \int_O f(t)\,dt - f^q(1). \tag{2.19}$$

Indeed, let $g(t) = f(xt)$. Then
$$\begin{aligned} Sg(x^n) &= g^q(x^{n-1}) + xg^q(x^{n-2}) + \cdots + x^{n-1}g^q(1) \\ &= f^q(x^n) + xf^q(x^{n-1}) + \cdots + x^{n-1}f^q(x) \\ &= (Sf)(x^{n+1}) - x^n f^q(1) \end{aligned}$$

whence
$$\frac{Sg(x^n)}{x^n} = x \cdot \frac{(Sf)(x^{n+1})}{x^{n+1}} - f^q(1),$$

and (2.19) is obtained by passing to the limit for $n \to \infty$.

Using (2.18) we obtain by induction that
$$\int_O f(x^n t)\,dt = x^n \int_O f(t)\,dt - x^{n-1}f^q(1) - x^{n-2}f^q(x) - \cdots - f^q(x^{n-1}).$$

This equality implies the following invariance property. Suppose that a function f vanishes on all elements $z \in \mathbb{F}_q[x]$ with $\deg z < n$. Then, if $g \in \mathbb{F}_q[x]$, $\deg g \leq n$, we have
$$\int_O f(gt)\,dt = g \int_O f(t)\,dt.$$

Our next result will contain the calculation of integrals for some important functions on O. For the definitions of the special functions used below see Chapter 1.

Theorem 2.6 (i) *For any* $n = 0, 1, 2, \ldots$

$$\int_O t^{q^n} \, dt = -\frac{1}{[n+1]}. \tag{2.20}$$

(ii) *For any* $n = 0, 1, 2, \ldots$

$$\int_O f_n(t) \, dt = \frac{(-1)^{n+1}}{L_{n+1}}. \tag{2.21}$$

(iii) *If* $z \in K$, $|z| < 1$, *then*

$$\int_O C_s(z) \, ds = \log_C(z) - z. \tag{2.22}$$

Proof. (i) We have seen that

$$d\left(\frac{t^{q^n}}{D_n}\right) = \frac{t^{q^{n-1}}}{D_{n-1}}, \quad n \geq 1.$$

It follows from the definition of the operator S that

$$S\left(\frac{t^{q^n}}{D_n}\right) = \frac{t^{q^{n+1}} - t}{D_{n+1}}$$

whence

$$S\left(t^{q^n}\right) = \frac{D_n^q}{D_{n+1}}\left(t^{q^{n+1}} - t\right) = \frac{t^{q^{n+1}} - t}{[n+1]},$$

$$\frac{S\left(t^{q^n}\right)(x^k)}{x^k} = \frac{x^{k(q^{n+1}-1)} - 1}{[n+1]} \longrightarrow -\frac{1}{[n+1]}, \quad k \to \infty.$$

(ii) We have

$$\int_O f_n(t) \, dt = (Sf_n)'(0) = f'_{n+1}(0).$$

According to (1.14), the linear term in the expression for f_i is

$$(-1)^i \frac{\begin{bmatrix} i \\ 0 \end{bmatrix}}{D_i} t = \frac{(-1)^i}{L_i} t. \qquad (2.23)$$

Differentiation yields (2.21).

(iii) Applying the operator d (with respect to the variable s) to the function $C_s(z)$ we find that

$$d_s C_s(z) = \sum_{i=1}^{\infty} f_{i-1}(s) z^{q^{i-1}} = C_s(z)$$

whence

$$S_s C_s(z) = C_s(z) - C_1(z)s = \sum_{i=0}^{\infty} f_i(s) z^{q^i} - zs.$$

Fixing z and denoting $\varphi(s) = S_s C_s(z)$ we obtain that

$$\frac{\varphi(x^n)}{x^n} = \sum_{i=0}^{\infty} \frac{f_i(x^n)}{x^n} z^{q^i} - z. \qquad (2.24)$$

It is seen from (1.14) and (2.23) that

$$\frac{f_i(x^n)}{x^n} \longrightarrow \frac{(-1)^i}{L_i} \quad \text{for } n \to \infty.$$

On the other hand, it follows from (2.6) and Proposition 1.17 that

$$\left| \frac{f_i(x^n)}{x^n} \right| \leq q^i$$

for all n. If $z \in O$, $|z| < 1$, then $|z| \leq q^{-1}$, and the series in (2.24) converges uniformly with respect to n. Passing to the limit $n \to \infty$ in (2.24) we come to (2.22). ■

Setting in (2.22) $z = e_C(t)$, $|t| < 1$, we get an identity for the Carlitz exponential:

$$\int_O e_C(st) \, ds = t - e_C(t).$$

On the other hand, the formula (2.22) implies a more general formula (conjectured by D. Goss).

Corollary 2.7 *If $a \in O$, $z \in K$, $|z| < 1$, then*

$$\int_O C_{sa}(z)\,ds = a\log_C(z) - C_a(z). \qquad (2.25)$$

Proof. Since it is easily shown that (for each fixed z) the mapping $O \to C_0^1(O, \overline{K}_c)$ of the form $a \mapsto C_{sa}(z)$ is continuous, it is sufficient to prove (2.25) for $a = x^n$, $n = 1, 2, \dots$ Using (2.22), we find that

$$\int_O C_{sx^n}(z)\,ds = x^n(\log_C(z) - z) - \sum_{k=1}^{n} x^{n-k} C^q_{x^{k-1}}(z). \qquad (2.26)$$

Let $t = \log_C(z)$. Then $z = e_C(t)$ (see Section 1.6.1). It follows from properties of e_C that

$$\sum_{k=1}^{n} x^{n-k} C^q_{x^{k-1}}(z) = \sum_{k=1}^{n} x^{n-k}(e_C(x^k t) - x e_C(x^{k-1} t))$$
$$= e_C(x^n t) - x^n e_C(t) = C_{x^n}(z) - x^n z.$$

Substituting this into (2.26) we come to (2.25). ■

As $C_{sa}(z) = C_s(C_a(z))$, equation (2.22) also implies that

$$\int_O C_{sa}(z)\,ds = \log_C(C_a(z)) - C_a(z).$$

Comparing this with (2.25) implies

$$a\log_C(z) = \log_C(C_a(z))$$

which is precisely the functional equation of $\log_C(z)$.

2.1.5. Fractional derivatives. In this section we introduce the operator $\Delta^{(\alpha)}$, $\alpha \in O$, a function field analog of the Hadamard fractional derivative $\left(t\frac{d}{dt}\right)^\alpha$ from real analysis (see [97]).

Let $\alpha \in O$, $\alpha = \sum_{n=0}^{\infty} \alpha_n x^n$, $\alpha_n \in \mathbb{F}_q$. Denote $\widehat{\alpha} = \sum_{n=0}^{\infty} (-1)^n \alpha_n x^n$. The transformation $\alpha \mapsto \widehat{\alpha}$ is an $\mathbb{F}_q$-linear isometry. For an arbitrary continuous $\mathbb{F}_q$-linear function u on O we define its "fractional derivative" $\Delta^{(\alpha)} u$ at a point $t \in O$ by the formula

$$\left(\Delta^{(\alpha)} u\right)(t) = \sum_{k=0}^{\infty} (-1)^k \mathcal{D}_k(\widehat{\alpha}) u(x^k t) \qquad (2.27)$$

where $\{\mathcal{D}_k(t)\}$ is the sequence of hyperdifferentiations (see Section 1.3). The series converges for each t, uniformly with respect to α, since $|\mathcal{D}_k(\widehat{\alpha})| \leq 1$ and $u(x^k t) \to 0$. Thus $\Delta^{(\alpha)} u$ is, for each t, a continuous $\mathbb{F}_q$-linear function in α.

Our understanding of $\Delta^{(\alpha)}$ as a kind of a fractional derivative is justified by the following result.

Proposition 2.8 $\Delta^{(x^n)} = \Delta^n$, $n = 1, 2, \ldots$

Proof. By the definition of $\mathcal{D}_k$, it follows from (2.27) that

$$\left(\Delta^{(x^n)} u\right)(t) = \sum_{k=0}^{n} \binom{n}{k} (-x)^{n-k} u(x^k t).$$

If $n = 1$, then $\left(\Delta^{(x)} u\right)(t) = u(xt) - x u(t) = (\Delta u)(t)$. Suppose we have proved that $\Delta^{(x^{n-1})} = \Delta^{n-1}$. Then

$$(\Delta^n u)(t) = \Delta\left(\Delta^{(x^{n-1})} u\right)(t)$$

$$= \sum_{k=0}^{n-1} \binom{n-1}{k} (-x)^{n-1-k} u(x^{k+1} t) - x \sum_{k=0}^{n-1} \binom{n-1}{k} (-x)^{n-1-k} u(x^k t)$$

$$= \sum_{k=1}^{n} \binom{n-1}{k-1} (-x)^{n-k} u(x^k t) + \sum_{k=0}^{n-1} \binom{n-1}{k} (-x)^{n-k} u(x^k t)$$

$$= u(x^n t) + \sum_{k=1}^{n-1} \left\{\binom{n-1}{k-1} + \binom{n-1}{k}\right\} (-x)^{n-k} u(x^k t) + (-x)^n u(t)$$

$$= \left(\Delta^{(x^n)} u\right)(t),$$

as desired. ∎

It follows from Proposition 2.8 that $\Delta^{(x^n)} \circ \Delta^{(x^m)} = \Delta^{(x^{n+m})} = \Delta^{(x^n \cdot x^m)}$, which prompts the following composition property.

Proposition 2.9 *For any $\alpha, \beta \in O$,*

$$\Delta^{(\alpha)} \left(\Delta^{(\beta)} u\right)(t) = \left(\Delta^{(\alpha\beta)} u\right)(t).$$

Proof. Using the Leibnitz rule (1.29) for hyperdifferentiations we have

$$\left(\Delta^{(\alpha)} \circ \Delta^{(\beta)} u\right)(t) = \sum_{k=0}^{\infty}(-1)^k \mathcal{D}_k(\widehat{\beta}) \sum_{l=0}^{\infty}(-1)^l \mathcal{D}_l(\widehat{\alpha}) u(x^{k+l} t)$$

$$= \sum_{n=0}^{\infty}(-1)^n u(x^n t) \sum_{k+l=n} \mathcal{D}_k(\widehat{\beta}) \mathcal{D}_l(\widehat{\alpha})$$

$$= \sum_{n=0}^{\infty}(-1)^n \mathcal{D}_n(\widehat{\alpha}\widehat{\beta}) u(x^n t)$$

$$= \left(\Delta^{(\alpha\beta)} u\right)(t). \blacksquare$$

2.2 Umbral calculus

2.2.1. Background. Classical umbral calculus [95, 92, 96, 91] is a set of algebraic tools for obtaining, in a unified way, a rich variety of results regarding the structure and properties of various polynomial sequences. There exists a lot of generalizations extending umbral methods to other classes of functions. However, there is a restriction common to the whole literature on umbral calculus – the underlying field must be of characteristic zero. An attempt to mimic the characteristic zero procedures in the positive characteristic case [37] revealed a number of pathological properties of the resulting structures. More importantly, these structures were not connected with the existing analysis in positive characteristic based on a completely different algebraic foundation.

Classically, the main notions of umbral calculus are the delta operator, which (like the derivative) decreases degrees of polynomials, and the sequence of binomial type, a polynomial sequence $\{p_n\}_{n=0}^{\infty}$, $\deg p_n = n$, such that

$$p_n(s+t) = \sum_{k=0}^{n} \binom{n}{k} p_k(s) p_{n-k}(t), \quad n = 0, 1, 2, \ldots \qquad (2.28)$$

Many classical polynomial systems satisfy this identity.

In order to guess the counterpart of (2.28) in $\mathbb{F}_q$-linear analysis over the field K, we look at the system of additive Carlitz polynomials $\{e_n\}$. Let us consider the Carlitz module function (1.68):

$$C_s(z) = \sum_{n=0}^{\infty} \frac{e_n(s)}{D_n} z^{q^n}, \quad |z| < 1. \qquad (2.29)$$

Calculus

If $s \in \mathbb{F}_q[x]$, then C_s is a polynomial in z, and (see (1.70))

$$C_{ts}(z) = C_t(C_s(z)), \quad s, t \in \mathbb{F}_q[x].$$

Let us write the last identity explicitly using (2.29). After rearranging the sums we find that

$$C_{ts}(z) = \sum_{i=0}^{\deg t + \deg s} z^{q^i} \sum_{\substack{m+n=i \\ m,n \geq 0}} \frac{1}{D_n D_m^{q^n}} e_n(t) \{e_m(s)\}^{q^n},$$

so that

$$e_i(st) = \sum_{n=0}^{i} \binom{i}{n}_K e_n(t) \{e_{i-n}(s)\}^{q^n} \tag{2.30}$$

where

$$\binom{i}{n}_K = \frac{D_i}{D_n D_{i-n}^{q^n}}. \tag{2.31}$$

In this section we show that the "K-binomial" relation (2.30), a positive characteristic counterpart of the classical binomial formula, can be used for developing umbral calculus in the spirit of [95]. In particular, we introduce and study corresponding (nonlinear) delta operators, obtain a representation for operators invariant with respect to multiplicative shifts, and construct generating functions for polynomial sequences of the K-binomial type. Such sequences are also used for constructing new orthonormal bases of the space $C_0(O, \overline{K}_c)$ (in particular, a sequence of the Laguerre-type polynomials), in a way similar to the p-adic (characteristic 0) case [113, 114, 90].

2.2.2. Delta operators and K-binomial sequences. Denote by $\overline{K}_c\{t\}$ the vector space over $\overline{K}_c$ consisting of $\mathbb{F}_q$-linear polynomials $u = \sum a_k t^{q^k}$ with coefficients from $\overline{K}_c$. We will often use the operator of multiplicative shift $(\rho_\lambda u)(t) = u(\lambda t)$ on $\overline{K}_c\{t\}$ and the Frobenius operator $\tau u = u^q$. We call a linear operator T on $\overline{K}_c\{t\}$ *invariant* if it commutes with ρ_λ for each $\lambda \in K$.

Lemma 2.10 *If T is an invariant operator, then $T(t^{q^n}) = c_n t^{q^n}$, $c_n \in \overline{K}_c$, for each $n \geq 0$.*

Proof. Suppose that

$$T(t^{q^n}) = \sum_{l=1}^{N} c_{j_l} t^{q^{j_l}}$$

where j_l are different nonnegative integers, $c_{j_l} \in \overline{K}_c$. For any $\lambda \in K$

$$\rho_\lambda T(t^{q^n}) = T\rho_\lambda(t^{q^n}) = T\left((\lambda t)^{q^n}\right) = \lambda^{q^n} T(t^{q^n}) = \lambda^{q^n} \sum_{l=1}^{N} c_{j_l} t^{q^{j_l}}.$$

On the other hand,

$$\rho_\lambda T(t^{q^n}) = \sum_{l=1}^{N} c_{j_l} \lambda^{q^{j_l}} t^{q^{j_l}}.$$

Since λ is arbitrary, this implies the required result. ■

If an invariant operator T is such that $T(t) = 0$, then by Lemma 2.10 the operator $\tau^{-1}T$ on $\overline{K}_c\{t\}$ is well-defined.

Definition 2.11 *An $\mathbb{F}_q$-linear operator $\delta = \tau^{-1}\delta_0$, where δ_0 is a linear invariant operator on $\overline{K}_c\{t\}$, is called a delta operator if $\delta_0(t) = 0$ and $\delta_0(f) \neq 0$ for $\deg f > 1$, that is $\delta_0(t^{q^n}) = c_n t^{q^n}$, $c_n \neq 0$, for all $n \geq 1$.*

The most important example of a delta operator is the Carlitz derivative $d = \tau^{-1}\Delta$,

$$(\Delta u)(t) = u(xt) - xu(t).$$

Definition 2.12 *A sequence $\{P_n\}_0^\infty$ of $\mathbb{F}_q$-linear polynomials is called a basic sequence corresponding to a delta operator $\delta = \tau^{-1}\delta_0$, if $\deg P_n = q^n$, $P_0(1) = 1$, $P_n(1) = 0$ for $n \geq 1$,*

$$\delta P_0 = 0, \quad \delta P_n = [n]^{1/q} P_{n-1}, \; n \geq 1, \tag{2.32}$$

or, equivalently,

$$\delta_0 P_0 = 0, \quad \delta_0 P_n = [n] P_{n-1}^q, \; n \geq 1. \tag{2.33}$$

It follows from well-known identities for the Carlitz polynomials e_i (see Proposition 1.7 and Theorem 1.24) that the sequence $\{e_i\}$ is basic with respect to the operator d. For the normalized Carlitz polynomials f_i we have the relations

$$df_0 = 0, \quad df_i = f_{i-1}, \; i \geq 1.$$

The next definition is a formalization of the property (2.30).

Definition 2.13 *A sequence of $\mathbb{F}_q$-linear polynomials $u_i \in \overline{K}_c\{t\}$ is called a sequence of K-binomial type if $\deg u_i = q^i$ and for all $i = 0, 1, 2, \ldots$*

$$u_i(st) = \sum_{n=0}^{i} \binom{i}{n}_K u_n(t) \{u_{i-n}(s)\}^{q^n}, \quad s, t \in K. \quad (2.34)$$

If $\{u_i\}$ is a sequence of K-binomial type, then $u_i(1) = 0$ for $i \geq 1$, $u_0(1) = 1$ (so that $u_0(t) = t$).

Indeed, for $i = 0$ the formula (2.34) gives $u_0(st) = u_0(s)u_0(t)$. Setting $s = 1$ we have $u_0(t) = u_0(1)u_0(t)$, and since $\deg u_0 = 1$, so that $u_0(t) \not\equiv 0$, we get $u_0(1) = 1$.

If $i > 0$, for all t

$$0 = u_i(t) - u_i(t) = \sum_{n=0}^{i-1} \binom{i}{n}_K \{u_{i-n}(1)\}^{q^n} u_n(t),$$

and the linear independence of the polynomials u_n means that $u_l(1) = 0$ for $l \geq 1$.

Theorem 2.14 *For any delta operator $\delta = \tau^{-1}\delta_0$, there exists a unique basic sequence $\{P_n\}$, which is a sequence of K-binomial type. Conversely, given a sequence $\{P_n\}$ of K-binomial type, define the action of δ_0 on P_n by the relations (2.33), extend it onto $\overline{K}_c\{t\}$ by linearity and set $\delta = \tau^{-1}\delta_0$. Then δ is a delta operator, and $\{P_n\}$ is the corresponding basic sequence.*

Proof. Let us construct a basic sequence corresponding to δ. Set $P_0(t) = t$ and suppose that P_{n-1} has been constructed. It follows from Lemma 2.10 that δ is surjective, and we can choose P_n satisfying (2.32). For any $c \in \overline{K}_c$, $P_n + ct$ also satisfies (2.32), and we may redefine P_n choosing c in such a way that $P_n(1) = 0$.

Hence, a basic sequence $\{P_n\}$ indeed exists. If there is another basic sequence $\{P'_n\}$ with the same delta operator, then $\delta(P_n - P'_n) = 0$, whence $P'_n(t) = P_n(t) + at$, $a \in \overline{K}_c$, and setting $t = 1$ we find that $a = 0$.

In order to prove the K-binomial property, we introduce some operators having an independent interest.

Consider the linear operators $\delta_0^{(l)} = \tau^l \delta^l$.

Lemma 2.15

(i) *The identity*

$$\delta_0^{(l)} P_j = \frac{D_j}{D_{j-l}^{q^l}} P_{j-l}^{q^l} \qquad (2.35)$$

holds for any $l \leq j$.

(ii) *Let f be an $\mathbb{F}_q$-linear polynomial, $\deg f \leq q^n$. Then a generalized Taylor formula*

$$f(st) = \sum_{l=0}^{n} \frac{\left(\delta_0^{(l)} f\right)(s)}{D_l} P_l(t) \qquad (2.36)$$

holds for any $s, t \in K$.

Proof. By (2.32),

$$\delta^l P_j = \delta^{l-1}\left([j]^{q^{-1}} P_{j-1}\right)$$
$$= [j]^{q^{-l}} \delta^{l-1} P_{j-1}$$
$$= [j]^{q^{-l}} [j-1]^{q^{-(l-1)}} \delta^{l-2} P_{j-2} = \ldots$$
$$= [j]^{q^{-l}} [j-1]^{q^{-(l-1)}} \ldots [j-(l-1)]^{q^{-1}} P_{j-l},$$

so that

$$\delta_0^{(l)} P_j = [j][j-1]^q \ldots [j-(l-1)]^{q^{l-1}} P_{j-l}^{q^l}$$

which is equivalent to (2.35).

Since $\deg P_j = q^j$, the polynomials $P_1, \ldots, P_n$ form a basis of the vector space of all $\mathbb{F}_q$-linear polynomials of degrees $\leq n$ (because its dimension equals n). Therefore

$$f(st) = \sum_{j=0}^{n} b_j(s) P_j(t) \qquad (2.37)$$

where $b_j(s)$ are, for each fixed s, some elements of $\overline{K}_c$.

Applying the operator $\delta_0^{(l)}$, $0 \leq l \leq n$, in the variable t to both sides of (2.37) and using (2.35) we find that

$$\left(\delta_0^{(l)} f\right)(st) = \sum_{j=l}^{n} b_j(s) \frac{D_j}{D_{j-l}^{q^l}} P_{j-l}^{q^l}(t)$$

(note also that $\delta_0^{(l)}$ commutes with ρ_s). Setting $t=1$ and taking into account that

$$P_{j-l}(1) = \begin{cases} 0, & \text{if } j > l; \\ 1, & \text{if } j = l, \end{cases}$$

we come to the equality

$$b_l(s) = \frac{\left(\delta_0^{(l)} f\right)(s)}{D_l}, \quad 0 \le l \le n,$$

which implies (2.36). ∎

Note that the formulas (2.35) and (2.36) for the Carlitz polynomials e_i were established long ago; see [43]. It is important that, in contrast to the classical umbral calculus, the linear operators involved in (2.36) are not powers of a single linear operator.

Proof of Theorem 2.14 (continued). In order to prove that $\{P_n\}$ is a sequence of K-binomial type, it suffices to take $f = P_n$ in (2.36) and to use the identity (2.35).

To prove the second part of the theorem, we calculate the action in the variable t of the operator δ_0, defined by (2.33), upon the function $P_n(st)$. Using the relation $D_{n+1} = [n+1]D_n^q$ we find that

$$\delta_{0,t} P_n(st) = \sum_{j=0}^{n} \binom{n}{j}_K P_{n-j}^{q^j}(s) \left(\delta_0 P_j\right)(t)$$

$$= \sum_{j=1}^{n} \frac{D_n}{D_{n-j}^{q^j} D_j} P_{n-j}^{q^j}(s)[j] P_{j-1}^q(t)$$

$$= \sum_{i=0}^{n-1} \frac{D_n[i+1]}{D_{n-i-1}^{q^{i+1}} D_{i+1}} P_{n-i-1}^{q^{i+1}}(s) P_i^q(t)$$

$$= [n] \sum_{i=0}^{n-1} \left(\frac{D_{n-1}}{D_{n-i-1}^q D_i}\right)^q P_{n-i-1}^{q^{i+1}}(s) P_i^q(t)$$

$$= [n] \left\{\sum_{i=0}^{n-1} \binom{n-1}{i}_K P_{n-i-1}^{q^i}(s) P_i(t)\right\}^q$$

$$= [n] P_{n-1}^q(st) = (\delta_0 P_n)(st),$$

that is, δ_0 commutes with multiplicative shifts.

It remains to prove that $\delta_0(f) \neq 0$ if $\deg f > 1$. Assuming that $\delta_0(f) = 0$ for $f = \sum_{j=0}^{n} a_j P_j$ we have

$$0 = \sum_{j=0}^{n} a_j[j] P_{j-1}^q = \left\{ \sum_{i=0}^{n-1} a_{i+1}^{1/q} [i+1]^{1/q} P_i \right\}^q$$

whence $a_1 = a_2 = \ldots = a_n = 0$ due to the linear independence of the sequence $\{P_i\}$. ∎

2.2.3. Invariant operators. Let T be a linear invariant operator on $\overline{K}_c\{t\}$. Let us find its representation via an arbitrary fixed delta operator $\delta = \tau^{-1}\delta_0$. By (2.36), for any $f \in \overline{K}_c\{t\}$, $\deg f = q^n$,

$$(Tf)(st) = (\rho_s T f)(t) = T(\rho_s f)(t) = T_t f(st) = \sum_{l=0}^{n} (TP_l)(t) \frac{\left(\delta_0^{(l)} f\right)(s)}{D_l}.$$

Setting $s = 1$ we find that

$$T = \sum_{l=0}^{\infty} \sigma_l \delta_0^{(l)} \qquad (2.38)$$

where $\sigma_l = \dfrac{(TP_l)(1)}{D_l}$. The infinite series in (2.38) actually becomes a finite sum if both sides of (2.38) are applied to any $\mathbb{F}_q$-linear polynomial $f \in \overline{K}_c\{t\}$.

Conversely, any such series defines a linear invariant operator on $\overline{K}_c\{t\}$.

Below we will consider in detail the case where δ is the Carlitz derivative d, so that $\delta_0 = \Delta$, and the operators $\delta_0^{(l)} = \Delta^{(l)}$ are given recursively:

$$\left(\Delta^{(l)} u\right)(t) = \left(\Delta^{(l-1)} u\right)(xt) - x^{q^{l-1}} u(t); \qquad (2.39)$$

see (1.18) (the formula (2.38) for this case was proved by a different method in [54]).

Using (2.39) with $l = 0$, we can compute for this case the coefficients c_n from Lemma 2.10. We have $\Delta^{(l)}(t^{q^n}) = 0$, if $n < l$,

$$\Delta(t^{q^n}) = [n] t^{q^n}, \quad n \geq 1;$$

$$\Delta^{(2)}(t^{q^n}) = \tau^2 d^2(t^{q^n}) = \tau \Delta \tau^{-1} \Delta(t^{q^n})$$
$$= \tau \Delta \left([n]^{1/q} t^{q^{n-1}}\right) = [n][n-1]^q t^{q^n}, \quad n \geq 2,$$

and by induction

$$\Delta^{(l)}(t^{q^n})[n][n-1]^q \ldots [n-l+1]^{q^{l-1}} t^{q^n} = \frac{D_n}{D_{n-l}^{q^l}} t^{q^n}, \quad n \geq l. \quad (2.40)$$

The explicit formula (2.40) makes it possible to find out when an operator $\theta = \tau^{-1}\theta_0$, with

$$\theta_0 = \sum_{l=1}^{\infty} \sigma_l \Delta^{(l)}, \quad (2.41)$$

is a delta operator. We have $\theta_0(t) = 0$,

$$\theta_0(t^{q^n}) = D_n S_n t^{q^n},$$

where $S_n = \sum_{l=1}^{n} \frac{\sigma_l}{D_{n-l}^{q^l}}$. Thus θ is a delta operator if and only if $S_n \neq 0$ for all $n = 1, 2, \ldots$

Example 1. Let $\sigma_l = 1$ for all $l \geq 1$, that is

$$\theta_0 = \sum_{l=1}^{\infty} \Delta^{(l)}. \quad (2.42)$$

Since $|D_i| = q^{-\frac{q^i-1}{q-1}}$, we have

$$\left| D_{n-l}^{q^l} \right| = q^{-\frac{q^n-q^l}{q-1}},$$

so that $|S_n| = q^{\frac{q^n-q}{q-1}}$ ($\neq 0$) by the ultra-metric property of the absolute value. Comparing (2.42) with a classical formula from [95] we may see the polynomials P_n for this case as analogs of the Laguerre polynomials.

Example 2. Let $\sigma_l = \frac{(-1)^{l+1}}{L_l}$. Now

$$S_n = \sum_{l=1}^{n} (-1)^{l+1} \frac{1}{L_l D_{n-l}^{q^l}}.$$

Let us use the identity

$$\sum_{j=0}^{h-1} \frac{(-1)^j}{L_j D_{h-j}^{q^j}} = \frac{(-1)^{h+1}}{L_h} \quad (2.43)$$

proved in [40]. It follows from (2.43) that

$$\sum_{j=1}^{h} \frac{(-1)^j}{L_j D_{h-j}^{q^j}} = \frac{(-1)^{h+1}}{L_h} - \frac{1}{D_h} + \frac{(-1)^h}{L_h} = -\frac{1}{D_h},$$

so that $S_n = D_n^{-1} (\neq 0)$, $n = 1, 2, \ldots$ In this case $\theta_0(t^{q^j}) = t^{q^j}$ for all $j \geq 1$ (of course, $\theta_0(t) = 0$), and $P_0(t) = t$, $P_n(t) = D_n \left(t^{q^n} - t^{q^{n-1}} \right)$ for $n \geq 1$.

2.2.4. Orthonormal bases. Let $\{P_n\}$ be the basic sequence corresponding to a delta operator $\delta = \tau^{-1}\delta_0$,

$$\delta_0 = \sum_{l=1}^{\infty} \sigma_l \Delta^{(l)}$$

(the operator series converges on any polynomial from $\overline{K}_c\{t\}$).
Let $Q_n = \dfrac{P_n}{D_n}$, $n = 0, 1, 2, \ldots$ Then for any $n \geq 1$

$$\delta Q_n = D_n^{-1/q} \delta P_n = \frac{[n]^{1/q}}{D_n^{1/q}} P_{n-1} = \frac{P_{n-1}}{D_{n-1}} = Q_{n-1},$$

and the K-binomial property of $\{P_n\}$ implies the identity

$$Q_i(st) = \sum_{n=0}^{i} Q_n(t) \left\{ Q_{i-n}(s) \right\}^{q^n}, \quad s, t \in K. \quad (2.44)$$

The identity (2.44) may be seen as another form of the K-binomial property. Though it resembles its classical counterpart, the presence of the Frobenius powers is a feature specific for the case of a positive characteristic. We will call $\{Q_n\}$ a *normalized basic sequence*.

Theorem 2.16 *If $|\sigma_1| = 1$, $|\sigma_l| \leq 1$ for $l \geq 2$, then the sequence $\{Q_n\}_0^{\infty}$ is an orthonormal basis of the space $C_0(O, \overline{K}_c)$ – for any $f \in C_0(O, \overline{K}_c)$ there is a uniformly convergent expansion*

$$f(t) = \sum_{n=0}^{\infty} \psi_n Q_n(t), \quad t \in O, \quad (2.45)$$

where $\psi_n = \left(\delta_0^{(n)} f \right)(1)$, $|\psi_n| \to 0$ as $n \to \infty$,

$$\|f\| = \sup_{n \geq 0} |\psi_n|. \quad (2.46)$$

Proof. We have $Q_0(t) = P_0(t) = t$, so that $\|Q_0\| = 1$. Let us prove that $\|Q_n\| = 1$ for all $n \geq 1$. Our reasoning will be based on expansions in the normalized Carlitz polynomials f_n.

Let $n = 1$. Since $\deg Q_n = q^n$, we have $Q_1 = a_0 f_0 + a_1 f_1$. We know that $Q_1(1) = f_1(1) = 0$, hence $a_0 = 0$, so that $Q_1 = a_1 f_1$. Next, $\delta Q_1 = Q_0 = f_0$. Writing this explicitly we find that

$$f_0 = a_1^{1/q} \tau^{-1} \sum_{l=1}^{\infty} \sigma_l \Delta^{(l)} f_1 = a_1^{1/q} \sigma_1^{1/q} d f_1 = a_1^{1/q} \sigma_1^{1/q} f_0,$$

whence $a_1 = \sigma_1^{-1}$, $Q_1 = \sigma_1^{-1} f_1$, and $\|Q_1\| = 1$.

Assume that $\|Q_{n-1}\| = 1$ and consider the expansion

$$Q_n = \sum_{j=1}^{n} a_j f_j$$

(the term containing f_0 is absent since $Q_n(1) = 0$). Applying δ we get

$$\delta Q_n = \sum_{j=1}^{n} a_j^{1/q} \sum_{l=1}^{\infty} \sigma_l^{1/q} \tau^{-1} \Delta^{(l)} f_j.$$

By (1.22),

$$\Delta^{(l)} e_j = \begin{cases} \frac{D_j}{D_{j-l}^{q^l}} e_{j-l}^{q^l}, & \text{if } l \leq j, \\ 0, & \text{if } l > j, \end{cases}$$

so that

$$\Delta^{(l)} f_j = \begin{cases} f_{j-l}^{q^l}, & \text{if } l \leq j, \\ 0, & \text{if } l > j. \end{cases}$$

Therefore

$$\delta Q_n = \sum_{j=1}^{n} a_j^{1/q} \sum_{l=1}^{j} \sigma_l^{1/q} f_{j-l}^{q^{l-1}}. \tag{2.47}$$

It follows from the identity $f_{i-1}^q = f_{i-1} + [i] f_i$ (a consequence of (1.21)) that

$$f_{j-l}^{q^{l-1}} = \sum_{k=0}^{l-1} \varphi_{j,l,k} f_{j-l+k}$$

where $\varphi_{j,l,0} = 1$, $|\varphi_{j,l,k}| < 1$ for $k \geq 1$. Substituting into (2.47) we find

that
$$Q_{n-1} = \sum_{j=1}^{n} a_j^{1/q} \sum_{l=1}^{j} \sigma_l^{1/q} \sum_{k=0}^{l-1} \varphi_{j,l,k} f_{j-l+k}$$
$$= \sum_{j=1}^{n} a_j^{1/q} \sum_{i=0}^{j-1} f_i \sum_{l=j-i}^{j} \sigma_l^{1/q} \varphi_{j,l,i-j+l}$$
$$= \sum_{i=0}^{n-1} f_i \sum_{j=i+1}^{n} a_j^{1/q} \sum_{l=j-i}^{j} \sigma_l^{1/q} \varphi_{j,l,i-j+l}$$

whence
$$\max_{0 \leq i \leq n-1} \left| \sum_{j=i+1}^{n} a_j^{1/q} \sum_{l=j-i}^{j} \sigma_l^{1/q} \varphi_{j,l,i-j+l} \right| = 1 \tag{2.48}$$

by the inductive assumption and the orthonormal basis property of the normalized Carlitz polynomials.

For $i = n-1$, we obtain from (2.48) that
$$\left| a_n^{1/q} \sum_{l=1}^{n} \sigma_l^{1/q} \varphi_{n,l,l-1} \right| \leq 1.$$

We have $\varphi_{n,1,0} = 1$, $|\sigma_1| = 1$, and
$$\left| \sum_{l=2}^{n} \sigma_l^{1/q} \varphi_{n,l,l-1} \right| < 1,$$

so that
$$\left| \sum_{l=1}^{n} \sigma_l^{1/q} \varphi_{n,l,l-1} \right| = 1$$

whence $|a_n| \leq 1$.

Next, for $i = n-2$ we find from (2.48) that
$$\left| a_{n-1}^{1/q} \sum_{l=1}^{n-1} \sigma_l^{1/q} \varphi_{n-1,l,l-1} + a_n^{1/q} \sum_{l=2}^{n} \sigma_l^{1/q} \varphi_{n,l,l-1} \right| \leq 1.$$

We have proved that the second summand on the left is in O; then the first summand is considered as above, so that $|a_{n-1}| \leq 1$. Repeating this reasoning we come to the conclusion that $|a_j| \leq 1$ for all j. Moreover, $|a_j| = 1$ for at least one value of j; otherwise we would come to a contradiction with (2.48). This means that $\|Q_n\| = 1$.

If f is an arbitrary $\mathbb{F}_q$-linear polynomial, $\deg f = q^N$, then by the generalized Taylor formula (2.36)

$$f(t) = \sum_{l=0}^{N} \psi_l Q_l(t), \quad t \in O,$$

where $\psi_l = \left(\delta_0^{(l)} f\right)(1)$.

Since $\|Q_l\| = 1$ for all l, we have $\|f\| \leq \sup_l |\psi_l|$. On the other hand, $\delta_0^{(l)} f = \tau^l \left(\tau^{-1} \delta_0\right)^l f$, and if we prove that $\|\delta_0 f\| \leq \|f\|$, this will imply the inequality $\|\delta_0^{(l)} f\| \leq \|f\|$. We have

$$\left\|\Delta^{(l)} f\right\| = \max_{t \in O} \left| \left(\Delta^{(l-1)} f\right)(xt) - x^{q^{l-1}} \left(\Delta^{(l-1)} f\right)(t) \right|$$
$$\leq \max_{t \in O} \left| \left(\Delta^{(l-1)} f\right)(t) \right| \leq \ldots \leq \max_{t \in O} |(\Delta f)(t)| \leq \|f\|,$$

so that

$$\|\delta_0 f\| = \left\| \sum_{l=0}^{\infty} \sigma_l \Delta^{(l)} f \right\| \leq \sup_l |\sigma_l| \cdot \left\|\Delta^{(l)} f\right\| \leq \|f\|$$

whence $\|\delta_0^{(l)} f\| \leq \|f\|$ and $\sup_l |\psi_l| \leq \|f\|$.

Thus, we have proved (2.46) for any polynomial. By a well-known result of non-Archimedean functional analysis (see Theorem 50.7 in [98]), the uniformly convergent expansion (2.45) and the equality (2.46) hold for any $f \in C_0(O, \overline{K}_c)$.

The relation $\psi_n = \left(\delta_0^{(n)} f\right)(1)$ also remains valid for any $f \in C_0(O, \overline{K}_c)$. Indeed, denote by $\varphi_n(f)$ a continuous linear functional on $C_0(O, \overline{K}_c)$ of the form $\left(\delta_0^{(n)} f\right)(1)$. Suppose that $\{F_N\}$ is a sequence of $\mathbb{F}_q$-linear polynomials uniformly convergent to f. Then

$$F_N = \sum_n \varphi_n(F_N) Q_n,$$

so that

$$F - F_N = \sum_{n=0}^{\infty} \{\psi_n - \varphi_n(F_N)\} Q_n,$$

and by (2.46),

$$\|F - F_N\| = \sup_n |\psi_n - \varphi_n(F_N)|.$$

For each fixed n we find that $|\psi_n - \varphi_n(F_N)| \leq \|F - F_N\|$, and passing to the limit as $N \to \infty$ we get that $\psi_n = \varphi_n(f)$, as desired. ∎

By Theorem 2.16, the Laguerre-type polynomial sequence from Example 1 is an orthonormal basis of $C_0(O, \overline{K}_c)$. The sequence from Example 2 does not satisfy the conditions of Theorem 2.16.

Note that the conditions $|\sigma_1| = 1$, $|\sigma_l| \leq 1$, $l = 2, 3, \ldots$, imply that $S_n \neq 0$ for all n, so that the series (2.41) considered in Theorem 2.16 always correspond to delta operators.

Let us write a recurrence formula for the coefficients of the polynomials Q_n. Here we assume only that $S_n \neq 0$ for all n. Let

$$Q_n(t) = \sum_{j=0}^{n} \gamma_j^{(n)} t^{q^j}. \qquad (2.49)$$

We know that $\gamma_0^{(0)} = 1$.

Using the relation $\delta_0\left(t^{q^n}\right) = D_n S_n t^{q^n}$ we find that for $n \geq 1$

$$Q_{n-1} = \delta Q_n = \tau^{-1} \sum_{j=1}^{n} \gamma_j^{(n)} D_j S_j t^{q^j} = \sum_{i=0}^{n-1} \left(\gamma_{i+1}^{(n)}\right)^{1/q} D_{i+1}^{1/q} S_{i+1}^{1/q} t^{q^i}.$$

Comparing this with the equality (2.49), with $n-1$ substituted for n, we get

$$\gamma_i^{(n-1)} = \left(\gamma_{i+1}^{(n)}\right)^{1/q} D_{i+1}^{1/q} S_{i+1}^{1/q}$$

whence

$$\gamma_{i+1}^{(n)} = \frac{\left(\gamma_i^{(n-1)}\right)^q}{D_{i+1} S_{i+1}}, \quad i = 0, 1, \ldots, n-1;\ n = 1, 2, \ldots \qquad (2.50)$$

The recurrence formula (2.50) determines all the coefficients $\gamma_i^{(n)}$ (if the polynomial Q_{n-1} is already known) except $\gamma_0^{(n)}$. The latter can be found from the condition $Q_n(1) = 0$:

$$\gamma_0^{(n)} = -\sum_{j=1}^{n} \gamma_j^{(n)}.$$

2.2.5. Generating functions. The definition (1.68) of the Carlitz module can be seen as a generating function for the normalized Carlitz polynomials f_i. Here we give a similar construction for the normalized basic sequence in the general case. As in Section 2.2.4, we consider a delta operator of the form $\delta = \tau^{-1}\delta_0$,

$$\delta_0 = \sum_{l=1}^{\infty} \sigma_l \Delta^{(l)}.$$

We assume that $S_n \neq 0$ for all n.

Let us define the generalized exponential

$$e_\delta(t) = \sum_{j=0}^{\infty} b_j t^{q^j} \qquad (2.51)$$

by the conditions $\delta e_\delta = e_\delta$, $b_0 = 1$. Substituting (2.51) we come to the recurrence relation

$$b_{j+1} = \frac{b_j^q}{D_{j+1} S_{j+1}} \qquad (2.52)$$

which determines e_δ as a formal power series.

Since $b_0 = 1$, the composition inverse $\log_\delta$ to the formal power series e_δ has a similar form:

$$\log_\delta(t) = \sum_{n=0}^{\infty} \beta_n t^{q^n}, \quad \beta_n \in K, \qquad (2.53)$$

(see Section 19.7 in [83] for a general treatment of formal power series of this kind). A formal substitution gives the relations

$$\beta_0 = 1, \quad \sum_{m+n=l} b_m \beta_n^{q^m} = 0, \quad l = 1, 2, \ldots,$$

whence

$$\beta_l = -\sum_{m=1}^{l} b_m \beta_{l-m}^{q^m}, \quad l = 1, 2, \ldots \qquad (2.54)$$

Theorem 2.17 *Suppose that $|\sigma_1| = 1$ and $|\sigma_l| \leq 1$ for all l. Then both the series (2.51) and (2.53) converge on the disk $D_q = \{t \in O : |t| \leq q^{-1}\}$, if $q \neq 2$, or $D_2 = \{t \in O : |t| \leq q^{-2}\}$, if $q = 2$, and*

$$e_\delta(t \log_\delta z) = \sum_{n=0}^{\infty} Q_n(t) z^{q^n}, \quad t \in O, \ z \in D_q. \qquad (2.55)$$

Proof. Since
$$\left|\frac{D_n}{D_{n-l}^{q^l}}\right| = q^{-\frac{q^l-1}{q-1}},$$

under our assumptions we have $|D_n S_n| = q^{-1}$ for all n. By (2.52), $|b_{j+1}| = q|b_j|^q$, $j = 0, 1, 2, \ldots$, and we prove easily by induction that

$$|b_j| = q^{\frac{q^j-1}{q-1}}, \quad j = 0, 1, 2, \ldots \tag{2.56}$$

For the sequence (2.54) we obtain the estimate

$$|\beta_j| \leq q^{\frac{q^j-1}{q-1}}, \quad j = 0, 1, 2, \ldots \tag{2.57}$$

Indeed, this is obvious for $j = 0$. If (2.57) is proved for $j \leq l-1$, then

$$|\beta_l| \leq \max_{1 \leq m \leq l} |b_m| \cdot |\beta_{l-m}|^{q^m} \leq \max_{1 \leq m \leq l} q^{\frac{q^m-1}{q-1} + q^m \frac{q^{l-m}-1}{q-1}} = q^{\frac{q^l-1}{q-1}}.$$

It follows from (2.56) and (2.57) that both the series (2.51) and (2.53) are convergent for $t \in D_q$ (in fact they are convergent on a wider disk from $\overline{K}_c$, but here we consider them only on K). Note also that

$$|\log_\delta(t)| \leq \max_{n \geq 0} q^{-\frac{1}{q-1}} \left(q^{\frac{1}{q-1}}|t|\right)^{q^n} = |t|$$

if $t \in D_q$.

If $\lambda \in D_q$, then the function $t \mapsto e_\delta(\lambda t)$ is continuous on O, and by Theorem 2.16

$$e_\delta(\lambda t) = \sum_{n=0}^{\infty} \psi_n(\lambda) Q_n(t)$$

where $\psi_n(\lambda) = \left(\delta_0^{(n)} e_\delta(\lambda \cdot)\right)(1) = \left(\delta_0^{(n)} e_\delta\right)(\lambda)$ due to the invariance of the operator $\delta_0^{(n)}$. Since $\delta_0^{(n)} = \tau^n \delta^n$ and $\delta e_\delta = e_\delta$, we find that $\delta_0^{(n)} e_\delta = e_\delta^{q^n}$. Therefore

$$e_\delta(\lambda t) = \sum_{n=0}^{\infty} Q_n(t) \{e_\delta(\lambda)\}^{q^n} \tag{2.58}$$

for any $t \in O$, $\lambda \in D_q$. Setting in (2.58) $\lambda = \log_\delta(z)$ we come to (2.55). ∎

2.3 Locally analytic functions

2.3.1. Interpolation series. The Mahler basis of $C(\mathbb{Z}_p, \mathbb{Q}_p)$ mentioned in Section 1.7.1 is a very special case of the general theory of interpolation series developed by Amice [3]. Here we give a brief description of its main results (for the case of a local field).

Let L be a non-Archimedean local field, and q_L be its residue field cardinality. We denote by O_L its ring of integers, and by π_L its prime element (see [59] for a summary of basic notions regarding local fields, as well as references for further reading). Let $M \subset O_L$ be an infinite compact subset; more generally, it would be possible to consider any compact subset of L of diameter ≤ 1. The distance function $d(t_1, t_2)$ on L induces a distance on M; for any $\alpha \in M$, denote by $V_k(\alpha)$ the ball in M of radius q_L^{-k}, $k \geq 0$, centered in α.

The compact set M is called *regular*, if every ball $V_{k-1}(\alpha)$ is a disjoint union of ν_k balls $V_k(\alpha_i)$, $i = 1, \ldots, \nu_k$. Note that for $M = O_L$, $\nu_k = q_L$ for any k.

Let us fix a sequence $\{t_n\} \subset M$. Consider the polynomials

$$P_n(t) = (t - t_0)(t - t_1) \cdots (t - t_{n-1}).$$

Suppose that the mapping $n \mapsto t_n$ is injective. Then we may introduce also the normalized polynomials

$$Q_n(t) = \frac{(t - t_0)(t - t_1) \cdots (t - t_{n-1})}{(t_n - t_0)(t_n - t_1) \cdots (t_n - t_{n-1})} = \frac{P_n(t)}{P_n(t_n)}.$$

Given a function $f \in C(M, L)$, we can introduce its interpolation polynomials $F_n \in L[t]$, $\deg F_n \leq n$, by the classical formulas

$$F_n(t) = \sum_{k=0}^{n} b_k Q_k(t), \quad b_k = \left(\sum_{j=0}^{k} \frac{f(t_j)}{P'_{k+1}(t_j)} \right) P_k(t_k),$$

equivalent to the Lagrange interpolation formula

$$F_n(t) = \left(\sum_{k=0}^{n} \frac{f(t_k)}{(t - t_k) P'_{n+1}(t_k)} \right) P_{n+1}(t).$$

In order to characterize those interpolation points $\{t_n\}$, for which the above polynomial systems possess basis properties, Amice introduced the notion of *a very well distributed sequence* $\{t_n\}$. Denote by M_k, $k \geq 0$, the quotient of M with respect to the equivalence relation: $\alpha \sim \beta$ $(\alpha, \beta \in M)$, if and only if $V_k(\alpha) = V_k(\beta)$. Let $\text{pr}_k : M \to M_k$ be the canonical projection. For an element $\mu \in M_i$ and $n \geq 1$ set

$$T_i(\mu) = \{k \in \mathbb{Z}_+ : \text{pr}_k(t_k) = \mu\},$$

$$\nu_i(\mu, n) = \text{card}(T_i(\mu) \cap \{0, 1, \ldots, n-1\}).$$

A sequence $\{t_n\}$ is called *well distributed* of order $h \in \mathbb{N}$, if, for any $i \leq h$,

$$\nu_i(\mu, n) \geq \text{int}\left(\frac{n}{\text{card } M_i}\right),$$

where $\text{int}(z)$ is the integral part of the real number z. Finally, $\{t_n\}$ is called *very well distributed* if it is well distributed of order h, for any $h \geq 1$.

The significance of this notion becomes clear from the following fundamental result.

Theorem 2.18 (Amice) *The polynomial sequence $\{Q_k\}$ is an orthonormal basis of the L-Banach space $C(M, L)$, if and only if the sequence $\{t_n\}$ is very well distributed.*

The above notions are useful also for the investigation of spaces of locally analytic functions. A function $f \in C(M, L)$ is called *locally analytic*, if for any $\mu \in M$ there exists a disk D_μ of nonzero radius containing μ, such that the restriction of f to $D_\mu \cap M$ can be extended to a function on D_μ represented by a convergent power series. More precisely, f is called locally analytic of order h ($h \in \mathbb{N}$) if the radius of the disk is greater than or equal to q_L^{-h}. The set of all such functions is denoted $A_h(M)$.

Suppose that M is a union of disks V_i of radius q^{-h}. Then the set $A_h(M)$ of locally analytic functions on M of order h can be made a Banach space with the norm $|f| = \sup_i |f_i|$ where f_i is a restriction of f to V_i, and $|f_i|$ is the supremum of the absolute values of the Taylor coefficients of f_i. Below we assume that $\text{diam } M = 1$.

Theorem 2.19 (i) *Let $h \in \mathbb{N}$. Suppose that elements $s_{n,h} \in L$ are chosen in such a way that the polynomials $R_{n,h} = \frac{1}{s_{n,h}} P_n$ have norm 1 in $A_h(M)$. The polynomials $R_{n,h}$ form an orthonormal basis of $A_h(M)$, if and only if* the sequence $\{t_n\}$ is well distributed of order h. In this case

$$|s_{n,h}| = q_L^{-\lambda_n}, \quad \lambda_n = \sum_{k=1}^{h} \operatorname{int}\left(\frac{n}{\operatorname{card} M_k}\right).$$

(ii) *The polynomial system $\{R_{n,h}\}$ is an orthonormal basis of $A_h(M)$ for any $h \geq 0$, if and only if the sequence $\{t_n\}$ is very well distributed.*

For the proofs of Theorems 2.18 and 2.19 see [3]. It is easy to reformulate Theorem 2.19 to obtain conditions for local analyticity of a function in terms of the coefficients of its expansion in the polynomial system $\{P_n\}$.

There are two major examples for Amice's theory. In the first example $L = \mathbb{Q}_p$, $M = \mathbb{Z}_p$, $t_n = n$, $n = 0, 1, 2, \ldots$ We get the Mahler basis in $C(\mathbb{Z}_p, \mathbb{Q}_p)$ and the local analyticity conditions in terms of the Mahler expansions. See [3] for the details.

In the second example, related to the material of this book, $L = K$, $M = O$, $\operatorname{card} M_k = q^k$. As shown by Wagner [118], a very well distributed sequence $\{t_n\}$ for this case can be constructed as follows. Let $S = \{\alpha_0, \alpha_1, \ldots, \alpha_{q-1}\} \subset O$ be a complete system of representatives of classes from O/P. For any nonnegative integer n, consider the digit expansion

$$n = n_0 + n_1 q + \cdots + n_s q^s, \quad 0 \leq n_i \leq q-1,$$

and set

$$t_n = \alpha_{n_0} + \alpha_{n_1} x + \cdots + \alpha_{n_s} x^s.$$

With this sequence, the polynomials Q_n form an orthonormal basis of the Banach space $C(M, K)$. The polynomials $\frac{1}{s_{n,h}} P_n$ with $s_{n,h} \in K$, form an orthonormal basis of $A_h(O)$, if and only if

$$|s_{n,h}| = q^{-\lambda_n}, \quad \lambda_n = \sum_{k=1}^{h} \operatorname{int}\left(\frac{n}{q^k}\right).$$

This result can be extended to the case of locally analytic functions with values in a Banach space E over K, in particular to the case $E = \overline{K}_c$.

2.3.2. The Carlitz expansions. Let us consider expansions of continuous locally analytic functions in the general Carlitz polynomials G_i (see Section 1.4.2).

Let h be a nonnegative integer. Denote
$$\mu_{n,h} = \sum_{i=h+1}^{\infty} \operatorname{int}\left(\frac{n}{q^i}\right); \quad n \in \mathbb{Z}_+.$$

Theorem 2.20 *The polynomials $x^{\mu_{n,h}} G_n(t)$, $n \geq 0$, form an orthonormal basis of the Banach space $A_h(O)$.*

Proof. Let us first consider relations between $\{Q_n\}$ and $\{G_n\}$ as orthonormal bases of $C(O, K)$. Both Q_n and G_n are polynomials of the same degree. Therefore for each $n \geq 0$, there exist elements $g_{nj} \in K$, $j = 0, 1, \ldots, n$, such that
$$G_n(t) = \sum_{j=0}^{n} g_{nj} Q_j(t), \quad \max_{0 \leq j \leq n} |g_{nj}| = 1.$$
Writing in matrix form, we get
$$\begin{pmatrix} G_0(t) \\ G_1(t) \\ \ldots \\ G_n(t) \end{pmatrix} = \begin{pmatrix} g_{00} & 0 & 0 & \ldots & 0 \\ g_{10} & g_{11} & 0 & \ldots & 0 \\ \ldots\ldots\ldots\ldots\ldots\ldots\ldots\ldots \\ g_{n0} & g_{n1} & g_{n2} & \ldots & g_{nn} \end{pmatrix} \begin{pmatrix} Q_0(t) \\ Q_1(t) \\ \ldots \\ Q_n(t) \end{pmatrix}.$$

By Theorem 2.19(i), $\{x^{\mu_{n,h}} Q_n(t)\}$ is an orthonormal basis of $A_h(O)$. For any $n \geq 0$, we write $R_n(t) = x^{\mu_{n,h}} Q_n(t)$, $H_n(t) = x^{\mu_{n,h}} G_n(t)$, and $\rho_{ij} = g_{ij} x^{\mu_{i,h} - \mu_{j,h}}$. We have
$$\begin{pmatrix} H_0(t) \\ H_1(t) \\ \ldots \\ H_n(t) \end{pmatrix} = \begin{pmatrix} \rho_{00} & 0 & 0 & \ldots & 0 \\ \rho_{10} & \rho_{11} & 0 & \ldots & 0 \\ \ldots\ldots\ldots\ldots\ldots\ldots\ldots\ldots \\ \rho_{n0} & \rho_{n1} & \rho_{n2} & \ldots & \rho_{nn} \end{pmatrix} \begin{pmatrix} R_0(t) \\ R_1(t) \\ \ldots \\ R_n(t) \end{pmatrix}.$$
where the diagonal elements ρ_{jj} have absolute value 1, while all other elements of the matrix (ρ_{ij}) belong to O, because
$$\mu_{i,h} - \mu_{j,h} = \sum_{i=h+1}^{\infty} \left(\operatorname{int}\left(\frac{i}{q^l}\right) - \operatorname{int}\left(\frac{j}{q^l}\right) \right) \geq 0,$$
if $i \geq j$. This proves that H_j belongs, for each j, to the unit ball of $A_h(O)$, and the reductions of these polynomials form a basis of the appropriate reduced $\mathbb{F}_q$-vector space. The desired result is a consequence of Proposition 1.5. ∎

Note that Theorem 2.20 remains valid for the space of functions with values in $\overline{K}_c$. It can be extended also to the case of a completion of $\mathbb{F}_q(x)$ at an arbitrary finite place; see [124].

Corollary 2.21 *Let* $f(t) = \sum_{n=0}^{\infty} a_n G_n(t)$, $a_n \in \overline{K}_c$, *be a continuous function on O, and let*

$$|a_n| = q^{-\alpha_n}, \quad \gamma = \liminf_{n \to \infty} \frac{\alpha_n}{n}.$$

Then:

(i) *f is locally analytic of order h, if and only if*

$$\alpha_n - \sum_{i=h+1}^{\infty} \operatorname{int}(n/q^i) \longrightarrow \infty, \text{ as } n \to \infty.$$

(ii) *f is locally analytic, if and only if $\gamma > 0$. If $\gamma > 0$ and*

$$l = \max(0, \operatorname{int}(-\log(q-1) + \log \gamma)/\log q) + 1),$$

then f is locally analytic of order $h \geq l$.

Proof. The first part follows directly from Theorem 2.20. In order to prove the second part, we give an estimate of the numbers $\mu_{n,h}$. Obviously,

$$\mu_{n,h} \leq \sum_{i=h+1}^{\infty} \frac{n}{q^i} = \frac{n}{(q-1)q^h}. \tag{2.59}$$

On the other hand, let us write the q-digit expansion

$$n = n_w q^w + \cdots + n_1 q + n_0, \quad n_w \neq 0.$$

Let $s(n, h) = n_w + \cdots + n_{h+2} + n_{h+1}$ and $t(n, h) = n_h q^h + \cdots + n_1 q + n_0$. Then

$$\operatorname{int}\left(\frac{n}{q^{h+1}}\right) = n_w q^{w-h-1} + \cdots + n_{h+2} q + n_{h+1},$$

$$\operatorname{int}\left(\frac{n}{q^{h+2}}\right) = n_w q^{w-h-2} + \cdots + n_{h+2},$$

$$\cdots \cdots$$

$$\operatorname{int}\left(\frac{n}{q^w}\right) = n_w.$$

Adding up these equalities and summing the progressions in the right-hand side, we get

$$\mu_{n,h} = n_w \frac{q^{w-h}-1}{q-1} + n_{w-1}\frac{q^{w-h-1}-1}{q-1} + \cdots + n_{h+1}\frac{q-1}{q-1}$$
$$= \frac{q^{-h}(n-t(n,h)) - s(n,h)}{q-1} = \frac{n - t(n,h) - q^h s(n,h)}{(q-1)q^h}.$$

It is easy to see that $s(n,h) \leq (q-1)(w-h) \leq (q-1)(\log_q n - h)$, $t(n,h) \leq (q-1)q^h$, whence

$$\mu_{n,h} \geq \frac{n}{(q-1)q^h} + h - \log_q n - 1. \qquad (2.60)$$

It follows from (2.59) and (2.60) that

$$\lim_{n\to\infty} \frac{\mu_{n,h}}{n} = \frac{1}{(q-1)q^h}. \qquad (2.61)$$

Now, since a locally analytic function is locally analytic of some order h, the equality (2.61) implies the equivalence of local analyticity and the relation $\gamma > 0$.

If $\gamma > 0$, then the inequality

$$\gamma - \frac{1}{(q-1)q^h} > 0$$

for a nonnegative integer h implies $f \in A_h(O)$. Solving the inequality with respect to h we come to the last assertion of our corollary. ∎

2.3.3. $\mathbb{F}_q$-linear functions. In applications, the result of Corollary 2.21 is often used for continuous $\mathbb{F}_q$-linear functions given by expansions in the normalized $\mathbb{F}_q$-linear Carlitz polynomials,

$$u(t) = \sum_{n=0}^{\infty} a_n f_n(t), \quad t \in O, \qquad (2.62)$$

$a_n \in \overline{K}_c$, $a_n \to 0$. For this case, the above result is formulated as follows.

Corollary 2.22 *The function u is locally analytic if and only if*

$$\gamma = \liminf_{n\to\infty}\{-q^{-n}\log_q |a_n|\} > 0, \qquad (2.63)$$

and if (2.63) holds, then u is analytic on any ball of radius q^{-l},

$$l = \max(0, \text{int}(-\log(q-1) + \log\gamma)/\log q) + 1).$$

Calculus

An important special case is that of $l = 0$, that is of functions (2.62), analytic on O. The result of Corollary 2.22 can be expressed in the following simpler form.

Corollary 2.23 *A function (2.62) is analytic on O, if and only if*

$$q^{q^n/(q-1)}|a_n| \longrightarrow 0, \quad \text{as } n \to \infty. \tag{2.64}$$

Let us give an elementary *Proof* of this fact, avoiding the use of Amice's theory.

Suppose that u is analytic on O, that is

$$u(t) = \sum_{n=0}^{\infty} b_n t^{q^n}, \quad \overline{K}_c \ni b_n \to 0. \tag{2.65}$$

By Theorem 1.8,

$$t^{q^n} = \sum_{j=0}^{n} \beta_{nj} e_j(t)$$

where

$$\beta_{nj} = \frac{\left(\Delta^{(j)} t^{q^n}\right)\big|_{t=1}}{D_j}.$$

We have seen in Section 1.7.2 (and used it in Section 2.1.1) that

$$\left(\sqrt[q^j]{\ } \circ \Delta^{(j)}\right)\left(\frac{t^{q^n}}{D_n}\right) = \frac{t^{q^{n-j}}}{D_{n-j}},$$

whence

$$\left(\Delta^{(j)} t^{q^n}\right)\big|_{t=1} = \frac{D_n}{D_{n-j}^{q^j}},$$

so that

$$t^{q^n} = \sum_{j=0}^{n} \frac{D_n}{D_{n-j}^{q^j}} f_j(t).$$

Substituting this into (2.65) and changing the order of summation we find that

$$a_n = \sum_{k=n}^{\infty} \frac{D_k}{D_{k-n}^{q^n}} b_k.$$

Since $|D_k| = q^{-(q^k-1)/(q-1)}$, we obtain that
$$|a_n| \leq q^{-(q^n-1)/(q-1)} \sup_{k \geq n} |b_k|,$$
which implies (2.64).

Conversely, (2.64) means that $a_n D_n \to 0$, as $n \to \infty$. Using the explicit formula (1.14) for the Carlitz polynomials we can write
$$u(t) = \sum_{n=0}^{\infty} \frac{a_n}{D_n} \sum_{j=0}^{n} (-1)^{n-j} \frac{D_n}{D_j L_{n-j}^{q^j}} t^{q^j}. \qquad (2.66)$$

It is easy to get that
$$\left| \frac{D_n}{D_j L_{n-j}^{q^j}} \right| = q^{s_{nj}}$$
where
$$s_{nj} = \begin{cases} (n-j)q^j - q^j - \cdots - q^{n-1}, & \text{if } n \geq j+1, \\ 0, & \text{if } n = j. \end{cases}$$

Since $s_{nj} \leq q^j \{(n-j-1) - q^{n-j-1}\}$, if $n \geq j+1$, and the function $z \mapsto q^z - z$ increases for $z \geq 1$, we have $s_{nj} < 0$. Now we may rewrite (2.66) in the form (2.65) with
$$|b_n| \leq \max_{k \geq n} \left| \frac{c_k}{D_k} \right| \longrightarrow 0, \quad n \to \infty. \qquad \blacksquare$$

2.4 General smooth functions

2.4.1. Smoothness in the sense of Schikhof. The notion of a smooth $\mathbb{F}_q$-linear function introduced in Section 2.1.2 is closely connected with the structures specific for $\mathbb{F}_q$-linear functions on subsets of a function field K – it involves the Carlitz higher difference operators. Unfortunately, no analog of the latter is known for general continuous functions on K or O. However, there exists a notion of smoothness proposed by Schikhof [98] for arbitrary non-Archimedean fields. As we will see, for our field K the smoothness property in the sense of Schikhof can be expressed equivalently in terms of coefficients of expansions in the general Carlitz polynomials.

For a positive integer n, set
$$\nabla^n O = \{(t_1, \ldots, t_n) \in O^n : t_i \neq t_j \text{ if } i \neq j\}.$$

The n-th order difference quotient $\Phi_n f : \nabla^{n+1} O \to K$ of a function $f : O \to K$ is inductively defined by $\Phi_0 f = f$,

$$(\Phi_n f)(t_1, t_2, \ldots, t_{n+1})$$
$$= (t_1 - t_2)^{-1} \{(\Phi_{n-1} f)(t_1, t_3, \ldots, t_{n+1}) - (\Phi_{n-1} f)(t_2, t_3, \ldots, t_{n+1})\}$$

for $n \geq 1$ and $(t_1, t_2, \ldots, t_{n+1}) \in \nabla^{n+1} O$. A function f is called a C^n-function ($f \in C^n(O, K)$), if and only if $\Phi_n f$ can be extended to a continuous function $\overline{\Phi_n f} : O^{n+1} \to K$.

For $f \in C^n(O, K)$ we define its n-th hyperderivative $D^{(n)} f$ setting

$$\left(D^{(n)} f\right)(t) = \overline{\Phi_n f}(t, t, \ldots t).$$

C^n-functions admit a characterization in terms of their Taylor expansions [98]. If $f \in C^n(O, K)$, then for all $t, y \in O$,

$$f(t) = f(y) + \sum_{j=1}^{n-1} (t-y)^j D^{(j)} f(y) + (t-y)^n \overline{\Phi_n f}(t, y, \ldots y)$$
$$= f(y) + \sum_{j=1}^{n} (t-y)^j D^{(j)} f(y) + (t-y)^n \left(\overline{\Phi_n f}(t, y, \ldots y) - D^{(n)} f(y)\right).$$

Conversely, let $f : O \to K$. If there exist continuous functions $\lambda_1, \ldots, \lambda_{n-1} : O \to K$ and $\Lambda_n : O \times O \to K$, such that

$$f(t) = f(y) + \sum_{j=1}^{n-1} (t-y)^j \lambda_j(y) + (t-y)^n \Lambda_n(t, y), \quad t, y \in O,$$

then $f \in C^n(O, K)$.

The above difference quotients can be written in another way. Let

$$\tilde{\nabla}^{n+1} O = \{(t, y_1, \ldots, y_n) \in O^{n+1} : y_j + \cdots + y_{j+k} \neq 0$$
$$\text{for } 1 \leq j \leq n \text{ and } 0 \leq k \leq n-j\}.$$

Define $\Psi_n f : \tilde{\nabla}^{n+1} O \to K$ by

$$(\Psi_n f)(t, y_1, \ldots, y_n) = (\Phi_n f)(y_1 + \cdots + y_n + t, y_1 + \cdots + y_{n-1} + t, \ldots, y_1 + t, t).$$

Then $f \in C^n(O, K)$ if and only if $\Psi_n f$ can be extended to a continuous function $\overline{\Psi_n f} : O^{n+1} \to K$.

The functions $\Psi_n f$ can also be defined inductively by $\Psi_0 f = f$,

$$(\Psi_n f)(t, y_1, \ldots, y_n)$$
$$= y_n^{-1} \left\{ (\Psi_{n-1} f)(t, y_1, \ldots, y_{n-1} + y_n) - (\Psi_{n-1} f)(t, y_1, \ldots, y_{n-1}) \right\},$$

$n \geq 1$.

2.4.2. Some auxiliary constructions. For any positive integer n, we write its q-digit expansion $n = n_0 + n_1 q + \cdots + n_w q^w$ with $n_w \neq 0$ and denote by $\nu(n)$ the largest integer ν, such that q^ν divides n. Let $l(n) = n_w q^w$. Let Γ_j be the factorial-like sequence defined by (1.4). Note that

$$\alpha_{\nu(n)} q^{\nu(n)} - 1 = \left(\alpha_{\nu(n)} - 1 \right) q^{\nu(n)} + (q-1) \left(q^{\nu(n)-1} + \cdots + 1 \right).$$

Using this identity together with Lemma 1.1(i) we find that

$$\frac{\Gamma_{n-1}}{\Gamma_n} = \frac{1}{L_{\nu(n)}}. \tag{2.67}$$

Define a new sequence of polynomials $\{H_n(t)\}$ setting $H_0(t) = 1$,

$$H_n(t) = \frac{\Gamma_{n+1} G_{n+1}(t)}{\Gamma_n t}, \quad n \geq 1.$$

Lemma 2.24 *The sequence $\{H_n(t)\}_{n \geq 0}$ is an orthonormal basis of $C(O, K)$.*

Proof. $\{H_n(t)\}$ is a polynomial of degree n, and its leading coefficient is Γ_n^{-1}, the same as the one for $G_n(t)$. It follows from (2.7) and (2.67) that

$$H_n(t) = g_{q^{\nu(n+1)}-1}(t) G_{n+1-q^{\nu(n+1)}}(t).$$

Taking into account Proposition 1.17 we see that $\|H_n\| \leq 1$, and the required property is (see the proof of Theorem 2.20) a consequence of Proposition 1.5. ∎

In fact, in the last proof we used the representation

$$H_n(t) = \sum_{j=0}^{\infty} \theta_{nj} G_j(t) \tag{2.68}$$

where $\theta_{nj} \in O$ for each $n, j \geq 0$, $\theta_{nj} = 0$ for $j > n$, and $|\theta_{nn}| = 1$ for each n. Inverting the triangular matrix of coefficients we find that

$$G_i(t) = \sum_{j=0}^{\infty} \gamma_{ij} H_j(t) \tag{2.69}$$

where $\gamma_{ij} \in O$, $|\gamma_{ii}| = 1$ for all i, j, $\gamma_{ij} = 0$ for $j > i$,

$$\sum_{k=0}^{\infty} \theta_{ik} \gamma_{kj} = \sum_{k=0}^{\infty} \gamma_{ik} \theta_{kj} = \delta_{ij}. \tag{2.70}$$

Lemma 2.25 *If a function $f : O \setminus \{0\} \to K$ can be expressed as a convergent series $f(t) = \sum\limits_{n=0}^{\infty} a_n H_n(t)$, for all $t \in O \setminus \{0\}$, then the sequence $\{a_n\}$ is uniquely determined by the values $f(t)$, $t \in O \setminus \{0\}$.*

Proof. Let $f(t) = 0$ for all $t \in O \setminus \{0\}$. We will show that $a_n = 0$ for any $n \geq 0$.

Let $m \in \mathbb{Z}_+$, $m < q^w - 1$, $w \in \mathbb{N}$. Denote by $S(w)$ the set of all nonzero polynomials from $\mathbb{F}_q[x]$ with degrees $< w$. From the definition of the polynomials H_n, changing the summation index we get

$$\sum_{n=1}^{q^w-1} a_{n-1} \frac{\Gamma_n}{\Gamma_{n-1}} \frac{G_n(\alpha)}{\alpha} = 0 \quad \text{for all } \alpha \in S(w).$$

Therefore

$$\sum_{\alpha \in S(w)} g_{q^w-1-(m+1)}(\alpha) \sum_{n=1}^{q^w-1} a_{n-1} \frac{\Gamma_n}{\Gamma_{n-1}} \frac{G_n(\alpha)}{\alpha} = 0,$$

so that

$$\sum_{n=1}^{q^w-1} a_{n-1} \frac{\Gamma_n}{\Gamma_{n-1}} \sum_{\deg \alpha < w} g_{q^w-1-(m+1)}(\alpha) G_n(\alpha) = 0,$$

since $G_n(0) = 0$ for any $n \geq 1$. Using Proposition 1.14 we find that $a_n = 0$ for all n. ∎

2.4.3. The Carlitz expansions. The main result of this section is as follows.

Theorem 2.26 *Let*

$$f(t) = \sum_{j=0}^{\infty} a_j G_j(t)$$

be a continuous function from O to K. Then $f \in C^m(O, K)$, if and only if

$$\lim_{j \to \infty} |a_j| j^m = 0. \tag{2.71}$$

Proof. Using the identities (1.43) and (2.67) we get, for $y_1 \neq 0$, that

$$\Psi_1 f(t, y_1) = \Phi_1 f(y_1 + t, t) = \frac{1}{y_1}(f(y_1 + t) - f(t))$$

$$= \sum_{n_0=0}^{\infty} a_{n_0} \frac{1}{y_1}(G_{n_0}(y_1 + t) - G_{n_0}(t))$$

$$= \sum_{n_0=0}^{\infty} a_{n_0} \sum_{j_1=1}^{n_0} \frac{1}{L_{\nu(j_1)}} H_{j_1-1}(y_1) \binom{n_0}{j_1} G_{n_0-j_1}(t)$$

$$= \sum_{j_1=1}^{\infty} \sum_{n_0=j_1}^{\infty} a_{n_0} \frac{1}{L_{\nu(j_1)}} \binom{n_0}{j_1} H_{j_1-1}(y_1) G_{n_0-j_1}(t)$$

$$= \sum_{n_0=0}^{\infty} \sum_{j_1=0}^{\infty} \binom{n_0+j_1+1}{j_1+1} \frac{a_{n_0+j_1+1}}{L_{\nu(j_1+1)}} H_{j_1}(y_1) G_{n_0}(t). \tag{2.72}$$

The above transformations are justified by the fact that in the last series the sequence of terms tends to zero, as $n_0 + j_1 \to \infty$, for any $y_1 \neq 0$, $t \in O$.

Sufficiency. Suppose that (2.71) holds. Since

$$\left|\frac{1}{L_{\nu(n)}}\right| \leq n, \tag{2.73}$$

the coefficients at $H_{j_1}(y_1)G_{n_0}(t)$ in (2.72) tend to 0, as $n_0 + j_1 \to \infty$. Let us rewrite (2.72) using (2.68). We find that

$$\Psi_1 f(t, y_1) = \sum_{n_0, n_1, j_1 \geq 0} \binom{n_0+j_1+1}{j_1+1} \frac{a_{n_0+j_1+1}}{L_{\nu(j_1+1)}} \theta_{j_1, n_1} G_{n_1}(y_1) G_{n_0}(t).$$

Using the recursion formula for the operations Ψ_m we obtain the expression

$$\Psi_m f(t, y_1, \ldots, y_m) = \sum_{n_{m-1}, \ldots, n_1, n_0, j_m \geq 0} b_{n_0, n_1, \ldots, n_{m-1}, j_m} H_{j_m}(y_m)$$

$$\times G_{n_{m-1}}(y_{m-1}) \ldots G_{n_1}(y_1) G_{n_0}(t), \tag{2.74}$$

where

$$b_{n_0,n_1,\ldots,n_{m-1},j_m}$$
$$= \sum_{j_1,\ldots,j_{m-1}\geq 0} \binom{n_0+j_1+1}{j_1+1}\binom{n_1+j_2+1}{j_2+1}\cdots\binom{n_{m-1}+j_m+1}{j_m+1}$$
$$\times \frac{a_{n_0+j_1+1}}{L_{\nu(j_1+1)}L_{\nu(j_2+1)}\cdots L_{\nu(j_m+1)}}\theta_{j_1,n_1+j_2+1}\cdots\theta_{j_{m-1},n_{m-1}+j_m+1}.$$
(2.75)

Since $\theta_{ij}=0$ for $i<j$, it follows from (2.73) that

$$\left|b_{n_0,n_1,\ldots,n_{m-1},j_m}\right| \leq |a_{n_0+j_1+1}|(n_0+j_1+1)^m.$$

In addition, we must have

$$j_1 \geq n_1+j_2+1,$$
$$j_2 \geq n_2+j_3+1,$$
$$\ldots\ldots\ldots\ldots$$
$$j_{m-1} \geq n_{m-1}+j_m+1,$$

if $b_{n_0,n_1,\ldots,n_{m-1},j_m} \neq 0$. Therefore

$$b_{n_0,n_1,\ldots,n_{m-1},j_m} \longrightarrow 0, \quad \text{as } n_0+n_1+\ldots+n_{m-1}+j_m \to \infty.$$

The sequence $\{H_{j_m}(y_m)G_{n_{m-1}}(y_{m-1})\ldots G_{n_1}(y_1)G_{n_0}(t)\}$, $j_m, n_{m-1}, \ldots, n_0 \geq 0$, is an orthonormal basis of $C(O^{m+1},K)$ (see [3]). Therefore $\Psi_m f$, thus $\Phi_m f$ can be extended to a continuous function from O^{m+1} to K, so that $f \in C^m(O,K)$.

Necessity. Let $f \in C^m(O,K)$. Then $f \in C^k(O,K)$ for $0 \leq k \leq m$. Therefore, in particular, $\Psi_1 f(t,y_1)$ can be extended to a continuous function $\overline{\Psi_1 f}$ from $O \times O$ to K. Since, as before, $\{H_{j_1}(y_1)G_{n_0}(y_0)\}_{j_1,n_0\geq 0}$ is an orthonormal basis of $C(O \times O, K)$, we have

$$\overline{\Psi_1 f}(t,y_1) = \sum_{n_0,j_1\geq 0} b_{n_0,j_1} H_{j_1}(y_1)G_{n_0}(t),$$

$b_{n_0,j_1} \to 0$, as $n_0+j_1 \to \infty$. Comparing this with (2.72) we find that

$$\sum_{n_0\geq 0}\sum_{j_1\geq 0}\left\{\binom{n_0+j_1+1}{j_1+1}\frac{a_{n_0+j_1+1}}{L_{\nu(j_1+1)}} - b_{n_0,j_1}\right\}H_{j_1}(y_1)G_{n_0}(t) = 0$$

for any $0 \neq y_1 \in O$, $t \in O$. By Lemma 2.25,

$$\binom{n_0+j_1+1}{j_1+1}\frac{a_{n_0+j_1+1}}{L_{\nu(j_1+1)}} = b_{n_0,j_1}, \quad n_0, j_1 \geq 0.$$

This equality means that the coefficients of $H_{j_1}(y_1)G_{n_0}(t)$ in (2.72) tend to 0, as $j_1 + n_0 \to \infty$. By (2.68), the continuation $\overline{\Psi_1 f}$ of $\Psi_1 f$ (for simplicity, now we preserve the notation $\Psi_1 f$) can be written as

$$\Psi_1 f(t, y_1) = \sum_{n_0,n_1,j_1 \geq 0} \binom{n_0+j_1+1}{j_1+1}\frac{a_{n_0+j_1+1}}{L_{\nu(j_1+1)}}\theta_{j_1,n_1} G_{n_1}(y_1) G_{n_0}(t).$$

Calculating inductively the further difference quotients, we obtain the expansion (2.74) with the coefficients $b_{n_0,n_1,\ldots,n_{m-1},j_m}$ of the form (2.75) and

$$b_{n_0,n_1,\ldots,n_{m-1},j_m} \longrightarrow 0, \quad \text{as } n_0 + n_1 + \ldots + n_{m-1} + j_m \to \infty.$$

It follows from (2.75) that

$$\sum_{j_1,\ldots,j_{m-1}} \binom{n_0+j_1+1}{j_1+1}\binom{n_1+j_2+1}{j_2+1}\cdots\binom{n_{m-1}+j_m+1}{j_m+1}$$

$$\times \frac{a_{n_0+j_1+1}}{L_{\nu(j_1+1)}L_{\nu(j_2+1)}\cdots L_{\nu(j_{m-1}+1)}}\theta_{j_1,n_1+j_2+1}\cdots\theta_{j_{m-1},n_{m-1}+j_m+1}$$

$$= L_{\nu(j_m+1)}b_{n_0,n_1,\ldots,n_{m-1},j_m}, \quad (2.76)$$

for all $j_m, n_0, n_1, \ldots, n_{m-1} \geq 0$. Let

$$i_{m-1} \geq 1,$$
$$n_{m-1} + j_m + 1 = k_{m-1} \quad (k_{m-1} \geq 1),$$
$$j_m + 1 = l(k_{m-1}).$$

By Lucas' theorem about binomial coefficients modulo a prime number (see [46]), in this case

$$\binom{n_{m-1}+j_m+1}{j_m+1} = \binom{k_{m-1}}{l(k_{m-1})} = 1.$$

Recalling (2.70) we multiply both sides of (2.76) by $\gamma_{k_{m-1},i_{m-1}}$ ($k_{m-1} = 1, 2, \ldots$), and add up to get

$$\sum_{j_1,\ldots,j_{m-2}} \binom{n_0 + j_1 + 1}{j_1 + 1} \cdots \binom{n_{m-3} + j_{m-2} + 1}{j_{m-2} + 1} \binom{n_{m-2} + j_{m-1} + 1}{i_{m-1} + 1}$$

$$\times \frac{a_{n_0+j_1+1}}{L_{\nu(j_1+1)} \cdots L_{\nu(i_{m-1}+1)}} \theta_{j_1,n_1+j_2+1} \cdots \theta_{j_{m-2},n_{m-2}+i_{m-1}+1}$$

$$= \sum_{\substack{k_{m-1} \geq 1 \\ j_m = l(k_{m-1}) - 1 \\ n_{m-1} = k_{m-1} - j_m - 1}} L_{\nu(j_m+1)} b_{n_0,n_1,\ldots,n_{m-1},j_m} \gamma_{k_m-1,i_m-1}$$

for all $i_{m-1} \geq 1$, and all $n_0, n_1, \ldots n_{m-2} \geq 0$. We change the index i_{m-1} to j_{m-1}, and keep doing the previous step on $j_{m-2}, \ldots, j_2, j_1$. After that we obtain the identity

$$\binom{n_0 + i_1 + 1}{i_1 + 1} \frac{a_{n_0+i_1+1}}{L_{\nu(i_1+1)}}$$

$$= \sum_{k_1,k_2,\ldots,k_{m-1}} L_{\nu(j_2+1)} \cdots L_{\nu(j_m+1)}$$

$$\times b_{n_0,n_1,\ldots,n_{m-1},j_m} \gamma_{k_m-1,i_m-1} \cdots \gamma_{k_2,j_2} \gamma_{k_1,i_1} \quad (2.77)$$

where the summation is made over those $k_{m-1} \geq 1$, $k_{m-2} \geq 2$, $\ldots$, $k_1 \geq m - 1$, for which $k_{m-1} \geq j_{m-1}, \ldots, k_2 \geq j_2$, $k_1 \geq i_1$, and $j_m = l(k_{m-1}) - 1$, $n_{m-1} = k_{m-1} - l(k_{m-1}), \ldots, j_2 = l(k_1) - 1$, $n_1 = k_1 - l(k_1)$.

Let us write the q-digit expansion $k = c_w q^w + c_{w-1} q^{w-1} + \cdots + c_0$, with $c_w \neq 0$ and w sufficiently large, and $i_1 + 1 = l(k) = c_w q^w$, $n_0 = k - l(k)$. Then $\nu(i_1 + 1) = w$, and we have inductively:

$$k_1 \geq i_1 \geq q^w - 1, \quad j_2 + 1 = l(k_1) \geq q^{w-1}, \quad \nu(j_2 + 1) \geq w - 1,$$
$$k_2 \geq j_2 \geq q^{w-1} - 1, \quad j_3 + 1 = l(k_2) \geq q^{w-2}, \quad \nu(j_3 + 1) \geq w - 2,$$
$$\cdots\cdots\cdots$$
$$k_{m-1} \geq j_{m-1} \geq q^{w-m+2} - 1, \quad j_m + 1 = l(k_{m-1}) \geq q^{w-m+1},$$
$$\nu(j_m + 1) \geq w - m + 1.$$

From the equality (2.77) we find that

$$a_k = L_w L_{w-1} \cdots L_{w-m+1} \sum_{k_1,\ldots,k_{m-1}} c_{k_1,\ldots,k_{m-1}} b_{n_0,n_1,\ldots,n_{m-1},j_m}$$

with $\left|c_{k_1,\ldots,k_{m-1}}\right| \leq 1$. We have $kq^{-1} < q^w \leq k$, so that

$$k^{-m} \leq |L_w L_{w-1} \cdots L_{w-m+1}| \leq k^{-m} C(m),$$

where $C(m)$ depends only on m. Therefore

$$|a_k|k^m = \Omega \left| \sum_{k_1,\ldots,k_{m-1}} c_{k_1,\ldots,k_{m-1}} b_{n_0,n_1,\ldots,n_{m-1},j_m} \right|$$

where Ω is a bounded sequence. As $j_m \to \infty$ when $k \to \infty$, we find that

$$\lim_{k\to\infty} |a_k|k^m = 0,$$

as desired. ∎

Theorem 2.25 remains valid for functions with values in $\overline{K}_c$. Comparing with Theorem 2.5 we see that $C_0^{k+1}(O, \overline{K}_c) \subset C^{q^k}(O, \overline{K}_c)$.

2.5 Entire functions

2.5.1. Definitions. In this section we consider functions on the field K_∞ (see Section 1.5) and the completion Ω of its algebraic closure $\overline{K_\infty}$. Just as in the construction of the field $\overline{K}_c$ used throughout this book, the absolute value $|\cdot|_\infty$ extends to an ultra-metric absolute value on Ω, and Ω is algebraically closed [98].

Below we often use the following notation. Let $0 \neq t \in \mathbb{F}_q(x)$; then $t = \dfrac{\theta_1}{\theta_2}$, $\theta_1, \theta_2 \in \mathbb{F}_q[x]$, and we write $|t|_\infty = q^{-\nu(t)}$ where $\nu(t) = \deg \theta_2 - \deg \theta_1$ is called *the valuation* on $\mathbb{F}_q(x)$. In particular, $\nu(t) = -\deg t$ for $t \in \mathbb{F}_q[x]$. As we identify K_∞ with the field $\mathbb{F}_q((x^{-1}))$ of formal power series in x^{-1}, we can extend ν onto K_∞ writing

$$\nu\left(\sum_{s=-\infty}^{\infty} a_s x^s\right) = -\sup\{r \in \mathbb{Z}, a_r \neq 0\}.$$

Note that $\nu(t) \geq 0$ if and only if $|t|_\infty \leq 1$; if $|t|_\infty = 1$, then $\nu(t) = 0$.

The relation $|t|_\infty = q^{-\nu(t)}$ is maintained for the extensions of $|\cdot|_\infty$ and ν onto Ω. On the other hand, we may call the number $\deg t = -\nu(t)$ the degree of an element $t \in \Omega$. We agree that $\deg 0 = -\infty$.

Let us consider a series

$$f(t) = \sum_{n=0}^{\infty} b_n t^n, \quad b_n \in \Omega. \tag{2.78}$$

For any real number r, denote

$$M_r(f) = \sup\{rn - \nu(b_n); \ n \in \mathbb{N}\} \tag{2.79}$$

where the supremum may take infinite values; for example, $M_r(0) = -\infty$. If $|z|_\infty \leq q^r$, then obviously $|f(t)|_\infty \leq q^{rn-\nu(b_n)}$, so that the function $M_r(f)$ defined in terms of the coefficients b_n describes the growth of the function f. For a precise description of this connection see Theorem 42.2 in [98].

As in classical analysis, a function (2.78) is called *entire* if the series in (2.78) converges everywhere on Ω. For a function f to be entire, it is necessary and sufficient that

$$\nu(b_n) + n\nu(t) \to \infty, \quad \text{as } n \to \infty, \tag{2.80}$$

for any $t \in \Omega$. Note that for an entire function f, $\nu(b_n)$ is positive for large values of n.

Denote by $\text{Ent}(\Omega)$ the set of all entire functions with coefficients in Ω. It follows from (2.80) that $M_r(f)$ is finite for any $f \in \text{Ent}(\Omega)$.

For $f \in \text{Ent}(\Omega)$, denote by $\mathcal{A}(q, f)$ the set of real numbers a possessing the following property: there exist real numbers $b = b(a) > 0$ and $r(a)$, such that for any $r \geq r(a)$, one has

$$M_r(f) \leq bq^{r^a}. \tag{2.81}$$

If the set $\mathcal{A}(q, f)$ is empty, we say that the *q-order* of f is $+\infty$; otherwise the lower bound $\omega(q, f)$ of the set $\mathcal{A}(q, f)$, possibly equal to $-\infty$, is called the *q-order* of f. Below we use for brevity the notation $\omega(f)$.

Suppose that $f \in \text{Ent}(\Omega)$, $\omega(f) \neq \pm\infty$. Denote by $\mathcal{B}(f)$ the set of real numbers $b \geq 0$ with the following property: there exists a real number $r(b)$ such that for any $r \geq r(b)$,

$$M_r(f) \leq bq^{r^{\omega(f)}}. \tag{2.82}$$

If $\mathcal{B}(f) = \emptyset$, we say that the *type* of f is $+\infty$; otherwise the type is the lower bound $\rho(f)$ of the set $\mathcal{B}(f)$.

2.5.2. The q-order and type of an entire function. We begin with an estimate for the coefficients b_n of an entire function (2.78).

Proposition 2.27 *Suppose that the series (2.78) defines an entire function on Ω with finite q-order. Then:*

(i) for any real number $\theta > \omega(f)$, there exists an integer $n_1 \in \mathbb{N}$ such that

$$\nu(b_n) \geq \frac{n \log_q n}{\theta}, \tag{2.83}$$

for any integer $n \geq n_1$;

(ii) *if $\omega(f) = 1$, and if f is of finite type, then for any $\alpha > \rho(f)$, there exists an integer $n_2 \in \mathbb{N}$ such that*

$$\nu(b_n) \geq n \log_q n - \frac{n}{\log q}(1 + \log(\alpha \log q)), \tag{2.84}$$

for any integer $n \geq n_2$.

Proof. Let $\omega(f) < \omega < \theta$. There exists $r_0 \in \mathbb{R}$ such that

$$\sup\{nr - \nu(b_n);\ n \geq 0\} \leq q^{r\omega}$$

for $r \geq r_0$. Let $n \geq q^{r_0 \omega}$, $r = \dfrac{\log n}{\omega \log q}$. Then

$$\nu(b_n) \geq \frac{n \log_q n}{\omega}\left(1 - \frac{\omega}{\log_q n}\right),$$

and taking

$$n \geq \max\left(q^{r_0 \omega}, q^{\frac{\theta \omega}{\theta - \omega}}\right)$$

we come to (2.83).

Suppose that $\omega(f) = 1$ and $\rho(f) < \infty$. Let $\alpha > \rho(f)$. There exists a real number r_1 such that

$$\sup\{nr - \nu(b_n);\ n \geq 0\} \leq \alpha q^r$$

for $r \geq r_1$. For $n \geq \alpha q^{r_1} \log q$, we get (2.84) taking

$$r = \log_q \frac{n}{\alpha \log q}. \qquad \blacksquare$$

The next result is an explicit description of the order and type of an entire function given by its coefficients b_n.

Proposition 2.28 *Let f be an entire function on Ω given by the series (2.78). Then*

$$\omega(f) = \limsup_{n \to \infty}\left(\frac{n \log_q n}{\nu(b_n)}\right). \tag{2.85}$$

If the q-order of f is finite, then

$$\rho(f) = \limsup_{r \to \infty}\left(\frac{M_r(f)}{q^{r\omega(f)}}\right). \tag{2.86}$$

Proof. Denote by $\lambda(f)$ the right-hand side of (2.85). If $\lambda(f) = +\infty$, then it follows from the definition that $\omega(f) = +\infty$. Suppose that $\lambda(f) < +\infty$. Let $\lambda > \lambda(f)$. Then there exists an integer n_1 such that

$$\frac{n \log_q n}{\nu(b_n)} < \lambda \quad \text{for } n \geq n_1.$$

Then $rn - \nu(b_n) < rn - \frac{1}{\lambda} n \log_q n$. Investigating (in an elementary way) the function $\varphi(n) = rn - \frac{1}{\lambda} n \log_q n$, we find that $\sup_n \varphi(n) \leq bq^{\lambda r}$, $b > 0$. This means that $\lambda \in \mathcal{A}(q, f)$, so that $\omega(f) \leq \lambda(f)$. The inverse inequality follows from (2.83). The proof of (2.86) is obvious. ∎

As an example of an entire function on Ω, consider the Carlitz exponential

$$e_C(t) = \sum_{n=0}^{\infty} \frac{t^{q^n}}{D_n}.$$

It follows from the definition (1.1) of the sequence $\{D_n\}$ that

$$\deg(D_n) = nq^n. \tag{2.87}$$

By (2.85) and (2.87), we find that $\omega(e_C) = 1$.
Next,

$$M_r(e_C) = \sup_{n \geq 1}(r - n)q^n.$$

Finding the maximum of the function $[1, \infty) \ni z \mapsto (r - z)q^z$ we see that it equals $\frac{1}{e \log q} q^r$ attained at $z = r - \frac{1}{\log q}$. By (2.86), this implies the equality $\rho(e_C) = \frac{1}{e \log q}$.

By the ultra-metric Liouville theorem, a bounded entire function over Ω is a constant ([98], Theorem 42.6). In this section we prove a K_∞-analog of a "more delicate" result by Pólya [84] which states that an entire function φ of (exponential) order less than 1 or order 1 and type less than $\log 2$, such that $\varphi(\mathbb{N}) \subset \mathbb{Z}$, is a polynomial.

First we formulate some auxiliary results, in fact, very interesting ones but lying beyond the scope of this book.

2.5.3. Factorials and interpolation.
A very general notion of a factorial, covering both the classical and Carlitz factorials, was proposed by Bhargava [13, 14].

Let $\mathfrak{K}$ be a local field with a prime element π and the ring of integers $\mathfrak{R}$ (in fact, Bhargava's construction deals with a more general framework of Dedekind domains), $S \subset \mathfrak{R}$ be an arbitrary subset. A π-*ordering* of S is a sequence $\Lambda = \{a_0, a_1, a_2, \ldots\} \subset S$ where a_0 is arbitrary and a_n is chosen recursively to minimize the valuation of $(a_n - a_0) \cdots (a_n - a_{n-1})$. The element

$$n!_\Lambda = (a_n - a_0) \cdots (a_n - a_{n-1})$$

called *the generalized factorial* generates the same ideal for any choice of Λ ([13], Theorem 1).

The n-th *generalized binomial polynomial* is then defined as

$$\binom{t}{n}_\Lambda = \frac{(t-a_0)\cdots(t-a_{n-1})}{n!_\Lambda}; \quad \binom{t}{0}_\Lambda = 1. \qquad (2.88)$$

By construction, $\binom{t}{n}_\Lambda$ maps S into $\mathfrak{R}$ for all $n \geq 0$. Moreover [13, 14], the generalized binomial polynomials (2.88) form an $\mathfrak{R}$-basis for the ring $\mathrm{Int}(S, \mathfrak{R})$ of polynomials over $\mathfrak{K}$ mapping S into $\mathfrak{R}$.

If $S = \mathfrak{R} = \mathbb{Z}_p$, then $\Lambda = \{0, 1, 2, \ldots\}$ is a p-ordering, and $n!_\Lambda = n!$ (in the classical sense). The above basis assertion in this case is close to Mahler's theorem about the binomial coefficient basis of $C(\mathbb{Z}_p, \mathbb{Z}_p)$ (see [98]).

There are some rings for which the above construction works in the "global", purely algebraic, setting. Such is, in particular, the ring $\mathbb{F}_q[x]$. Define a one-to-one correspondence between $\mathbb{N}$ and $\mathbb{F}_q[x]$ as follows. For every $n \in \mathbb{N}$, let $n = \sum_{i=0}^{s} n_i q^i$ be its q-adic expansion. Then, put

$$u_n = \sum_{i=0}^{s} u_{n_i} x^i$$

where $u_0, u_1, \ldots, u_{q-1}$ are the elements of $\mathbb{F}_q$. This sequence can be interpreted as a kind of ordering [13, 14] defining Bhargava's factorial

$$n!_{\mathbb{F}_q[x]} = \prod_{k=0}^{n-1}(u_n - u_k). \qquad (2.89)$$

It can be shown that, up to nonzero factors from $\mathbb{F}_q$, the elements (2.89) coincide with Carlitz's general factorials

$$\Gamma_i = \prod_{i=0}^{s} D_i^{n_i}$$

(see Chapter 1). The sequence of polynomials

$$\binom{t}{n}_{\mathbb{F}_q[x]} = \Gamma_n^{-1} \prod_{k=0}^{n-1}(t - u_k) \qquad (2.90)$$

is a basis of the $\mathbb{F}_q$-module $\mathrm{Int}(\mathbb{F}_q[x])$ of polynomials over $\mathbb{F}_q(x)$ taking values from $\mathbb{F}_q[x]$ on elements from $\mathbb{F}_q[x]$.

Note that the polynomial $\binom{t}{q^m}_{\mathbb{F}_q[x]}$, $m \geq 0$, coincides with the normalized Carlitz polynomial $f_m(t)$.

A detailed study of the polynomials (2.90) is given in [1]. Here we present some of the results without proofs.

Define the elements $a_{n,k}$ and $b_{n,k}$ of $\mathbb{F}_q[x]$ $(n, k \in \mathbb{N})$ by

$$t^n = \sum_{k=0}^{n} b_{n,k} \binom{t}{n}_{\mathbb{F}_q[x]},$$

$$\Gamma_n \binom{t}{n}_{\mathbb{F}_q[x]} = \sum_{k=0}^{n}(-1)^{n-k} a_{n,k} t^k.$$

We have $b_{n,k} = a_{n,k} = 0$ for $k > n$ and $k < 0$. We see also that $b_{n,0} = a_{n,0} = 0$ for $n > 0$.

Lemma 2.29 *The coefficients $b_{n,k}$ and $a_{n,k}$ satisfy the recursive relations*

$$a_{n+1,k} = u_n a_{n,k} + a_{n,k-1},$$

$$b_{n,k} = L_{e(k)} b_{n-1,k-1} + u_k b_{n-1,k},$$

where $e(k)$ denotes the highest power of q dividing k.

The above recursive relations are used to obtain estimates of the coefficients $a_{n,k}, b_{n,k}$.

Proposition 2.30 *If $1 \leq k \leq n$, then*

$$\deg\left(\frac{a_{n,k}}{\Gamma_n}\right) \leq -\log_q n + \frac{2q-1}{q-1} k - k \log_q k; \qquad (2.91)$$

$$\deg b_{n,k} \leq n \log_q k. \qquad (2.92)$$

It follows from (2.91) that values of the polynomial $\binom{t}{n}_{\mathbb{F}_q[x]}$ satisfy the estimate

$$\deg \binom{t}{n}_{\mathbb{F}_q[x]} \leq -\log_q n + \frac{q^{\frac{2q-1}{q-1}+\delta}}{e \log q}, \quad n \geq 1, \qquad (2.93)$$

if $x \in \Omega$ is of degree δ.

2.5.4. Interpolation series for entire functions. Consider a polynomial $g \in \Omega[t]$ of degree k,

$$g(t) = \sum_{n=0}^{k} c_n t^n.$$

We have

$$g(t) = \sum_{n=0}^{k} c_n \sum_{j=0}^{n} b_{n,j} \binom{t}{j}_{\mathbb{F}_q[x]},$$

so that

$$g(t) = \sum_{j \geq 0} \Delta_j(g) \binom{t}{j}_{\mathbb{F}_q[x]}$$

where

$$\Delta_j(g) = \sum_{n \geq j} c_n b_{n,j}. \qquad (2.94)$$

Let $f(t) = \sum_{n=0}^{\infty} c_n t^n$ be an entire function on Ω. If $j \geq 0$ and $\lim_{n \to \infty} \deg(c_n b_{n,j}) = -\infty$, we put, extending (2.94),

$$\Delta_j(f) = \sum_{n=j}^{\infty} c_n b_{n,j}.$$

Theorem 2.31 *Let $f \in \mathrm{Ent}(\Omega)$ and $\omega(f) < 1$, or $\omega(f) = 1$ and $\rho(f) < \dfrac{1}{e \log q}$. Then for all $j \in \mathbb{N}$, $\Delta_j(f)$ exists, and for all $t \in \Omega$,*

$$f(t) = \sum_{j=0}^{\infty} \Delta_j(f) \binom{t}{j}_{\mathbb{F}_q[x]}. \qquad (2.95)$$

Calculus

Proof. The assumption can be rephrased as $\tau(f) < \dfrac{1}{e \log q}$ where

$$\tau(f) = \limsup_{r \to \infty} \frac{M_r(f)}{q^r}.$$

Let $\tau > 0$ be such that $\tau(f) < \tau < \dfrac{1}{e \log q}$. By Proposition 2.27,

$$\deg c_n \leq n\theta - n \log_q n, \quad n \geq N_1,$$

where $\theta = \log_q(e\tau \log q) < 0$. Suppose that $n \geq j$. It follows from (2.92) that

$$\deg(c_n b_{n,j}) \leq n\theta - n \log_q n + n \log_q j \to -\infty,$$

as $n \to \infty$. This proves the existence of $\Delta_j(f)$. An elementary estimate yields the bound

$$\deg \Delta_j(f) \leq \theta j, \quad j \geq N_1. \tag{2.96}$$

Let $t \in \Omega$ be of degree δ. By (2.93) and (2.96),

$$\deg \left(\Delta_j(f) \binom{t}{j} \right)_{\mathbb{F}_q[x]} \leq -\log_q j + \frac{q^{\frac{2q-1}{q-1}+\delta}}{e \log q} + \theta j \to -\infty,$$

as $j \to \infty$. This implies the convergence of the series in the right-hand side of (2.95).

Denote

$$\overline{f}(t) = \sum_{j=0}^{\infty} \Delta_j(f) \binom{t}{j}_{\mathbb{F}_q[x]},$$

$$f_N(t) = \sum_{n=0}^{N-1} c_n t^n, \quad \overline{f}_N(t) = \sum_{j=0}^{N-1} \Delta_j(f) \binom{t}{j}_{\mathbb{F}_q[x]}.$$

Suppose that $t \in \Omega$ is of degree δ. Let $A > 0$. Since $f_N(t) \to f(t)$, as $N \to \infty$, there exists a natural number N_2 such that

$$\deg(f(t) - f_N(t)) \leq -A, \quad N \geq N_2. \tag{2.97}$$

There exists also $N_3 \in \mathbb{N}$, such that

$$\deg(\overline{f}(t) - \overline{f}_N(t)) \leq -A, \quad N \geq N_3. \tag{2.98}$$

Considering $N \geq N_1$ we have

$$f_N(t) - \overline{f}_N(t) = \sum_{j=0}^{N-1} [\Delta_j(f_N) - \Delta_j(f)] \binom{t}{j}_{\mathbb{F}_q[x]}$$

where $\Delta_0(f_N) = \Delta_0(f)$ and

$$\Delta_j(f_N) - \Delta_j(f) = \sum_{n \geq N} c_n b_{n,j}.$$

For all $n \geq N$, $\deg(c_n b_{n,j}) \leq n(\theta + \log_q j) - n\log_q n \leq N\theta$. Therefore

$$\deg(\Delta_j(f_N) - \Delta_j(f)) \leq N\theta,$$

$$\deg\left((\Delta_j(f_N) - \Delta_j(f))\binom{t}{j}_{\mathbb{F}_q[x]}\right) \leq N\theta - \log_q j + \frac{q^{\frac{2q-1}{q-1}+\delta}}{e\log q},$$

$$\deg(f_N(t) - \overline{f}_N(t)) \leq N\theta + \frac{q^{\frac{2q-1}{q-1}+\delta}}{e\log q}.$$

This results in the existence of $N_4 \in \mathbb{N}$ such that

$$\deg(f_N(t) - \overline{f}_N(t)) \leq -A, \quad N \geq N_4. \tag{2.99}$$

It follows from (2.97), (2.98), and (2.99) that $\deg(f(t) - \overline{f}(t)) \leq -A$. Since A is arbitrary, $f(t) = \overline{f}(t)$, which means the required equality (2.95). ∎

The analog of Pólya's theorem is as follows.

Theorem 2.32 *Let $f \in \mathrm{Ent}(\Omega)$ satisfy the conditions of Theorem 2.3.1 and, in addition, $f(\mathbb{F}_q[x]) \subset \mathbb{F}_q[x]$. Then f is a polynomial.*

Proof. By Theorem 2.31, for any $t \in \Omega$ we have the representation (2.95). Substituting $t = u_n$, $n = 0, 1, 2, \ldots$, we conclude that the sequence $\{\Delta_j(f)\}$ can be seen as a solution of the linear system

$$f(u_0) = \Delta_0(f);$$

$$f(u_1) = \Delta_0(f)\binom{u_1}{0}_{\mathbb{F}_q[x]} + \Delta_1(f)\binom{u_1}{1}_{\mathbb{F}_q[x]},$$

$$\ldots\ldots\ldots\ldots\ldots\ldots\ldots\ldots\ldots\ldots\ldots\ldots$$

$$f(u_n) = \Delta_0(f)\binom{u_n}{0}_{\mathbb{F}_q[x]} + \Delta_1(f)\binom{u_n}{1}_{\mathbb{F}_q[x]} + \cdots + \Delta_n(f)\binom{u_n}{n}_{\mathbb{F}_q[x]}$$

$$\ldots\ldots\ldots\ldots\ldots\ldots\ldots\ldots\ldots\ldots\ldots\ldots$$

For all $n \in \mathbb{N}$, $f(u_n) \in \mathbb{F}_q[x]$ (by our assumption), $\binom{u_n}{n}_{\mathbb{F}_q[x]} \in \mathbb{F}_q^*$ (by the relation between Bhargava's and Carlitz's factorials), and $\binom{u_n}{j}_{\mathbb{F}_q[x]} \in \mathbb{F}_q[x]$ for all $j < n$ (since the polynomials (2.90) belong to $\text{Int}(\mathbb{F}_q[x])$). By induction, we deduce that $\Delta_j(f) \in \mathbb{F}_q[x]$ for all $j \in \mathbb{N}$. Meanwhile we know that $\deg \Delta_j(f) \leq \theta j$, for j large enough. Since $\theta < 0$, this means that $\Delta_j(f) = 0$ for j large enough, so that f is a polynomial. ∎

It can be shown [21] that the function

$$f(t) = \sum_{n=0}^{\infty} \binom{t}{q^n}_{\mathbb{F}_q[x]} = \sum_{n=0}^{\infty} f_n(t)$$

is an entire function on Ω, $f(\mathbb{F}_q[x]) \subset \mathbb{F}_q[x]$, $\omega(f) = 1$, $\rho(f) = \frac{1}{e \log q}$, and, of course, f is not a polynomial. Thus, the bound $\frac{1}{e \log q}$ in Theorem 2.32 cannot be improved.

2.6 Measures and divided power series

2.6.1. Non-Archimedean measures. Let K_π be a completion of the field $\mathbb{F}_q(x)$ at a finite place, corresponding to an irreducible polynomial π (see Section 1.5). As usual, we denote by O_π the ring of integers in K_π. Let H be a compact totally disconnected Abelian topological group. It is well known (see e.g. [58]) that H has a fundamental system of subgroup neighborhoods $\{H_j\}_{j=1}^{\infty}$ of the identity e, such that

$$H_{j+1} \subset H_j, \quad \bigcap_{j=1}^{\infty} H_j = \{e\}, \quad H/H_j \text{ is finite,}$$

and H can be represented as a projective limit

$$H = \varprojlim H/H_j.$$

A π-adic measure on H is a finitely additive O_π-valued function μ on the algebra of compact open subsets of H. Let $\text{pr}_j : H \to H/H_j$ be the canonical mapping. For any $h \in H/H_n$, the inverse image $\text{pr}_n^{-1}(h)$ is open in H, and the collection of all such sets forms a base of the projective limit topology in H. If φ is a function from H/H_j to O_π, and $f : H \to O_\pi$ is a function of the form $f(h) = \varphi(\text{pr}_j h)$, $h \in H$, then the function f is called *locally constant* (obviously, f is constant on a neighborhood of each point).

For any locally constant function f, we can define its integral as

$$\int_H f(h)\,d\mu(h) = \sum_{t \in H/H_j} \varphi(t)\mu(\mathrm{pr}_j^{-1}(t)),$$

and this definition does not depend on the number j admissible in the definition of local constancy.

More generally, a continuous O_π-valued function on H can be approximated uniformly by locally constant functions. Then an obvious approximation procedure leads to a definition of the integral $\int_H f(h)\,d\mu(h)$ for any continuous function. This integral establishes the duality of the space of measures to the space of continuous functions.

It is important that the measure $\mu(S)$ is bounded in absolute value by a constant independent of the compact open set S. In our case this constant equals 1, since we have assumed that μ is O_π-valued. We omit the details of the above integration theory, since they are given in any textbook on non-Archimedean analysis; see, for example, [98]. Note however that the Volkenborn-type integration of Section 2.1.4 is not covered by the above scheme (it can actually be interpreted as integration with respect to an unbounded measure).

Let $M[H]$ be the O_π-module of O_π-valued measures on H. Then $M[H]$ can be furnished with the convolution product $\mu * \nu$ defined by the relation

$$\int_H f(h)\,d(\mu * \nu)(h) = \int_H \int_H f(h+s)\,d\mu(h)d\nu(s). \tag{2.100}$$

At this point we specialize to the case $H = O_\pi$ (see [43, 44] for the case $H = \mathbb{Z}_p$). As we know (Theorem 1.21), any continuous function $f: O_\pi \to O_\pi$ can be written uniquely as

$$f(t) = \sum_{j=0}^{\infty} a_j G_j(t), \quad O_\pi \ni a_j \to 0.$$

Here G_j are the general Carlitz polynomials. Thus an O_π-valued measure μ is uniquely determined by the sequence

$$b_j = \int_{O_\pi} G_j(t)\,d\mu(t). \tag{2.101}$$

This implies the following result.

Theorem 2.33 *An O_π-valued measure μ on O_π is uniquely determined by its moments*

$$\int_{O_\pi} t^i \, d\mu(t), \quad i = 0, 1, 2, \ldots$$

2.6.2. Divided power series. Given an O_π-valued measure μ on O_π, we associate to it the *formal divided power series*

$$P_\mu(z) = \sum_{j=0}^{\infty} b_j \frac{z^j}{j!} \tag{2.102}$$

where b_j is given by (2.101), and $\dfrac{z^j}{j!}$ is a formal symbol which must not be understood literally (since $j! = 0$ in O_π for $j \geq p$). However the multiplication of such symbols is defined in a natural way:

$$\frac{z^j}{j!} * \frac{z^k}{k!} = \binom{j+k}{j} \frac{z^{j+k}}{(j+k)!}.$$

Defining the addition in a termwise manner and assuming the distributive property we make the set of all formal divided power series a commutative ring. Thus, if we have two formal divided power series,

$$P(z) = \sum_{j=0}^{\infty} b_j^{(1)} \frac{z^j}{j!}, \quad Q(z) = \sum_{j=0}^{\infty} b_j^{(2)} \frac{z^j}{j!},$$

then

$$(P * Q)(z) = \sum_{j=0}^{\infty} \left(\sum_{k+l=j} \binom{k+l}{k} b_k^{(1)} b_l^{(2)} \right) \frac{z^j}{j!}. \tag{2.103}$$

Comparing (2.103) with an expression for the formal divided power series corresponding to the convolution (2.100) (here we use the identity (1.43)) we see that the mapping

$$\mu \mapsto \sum_{j=0}^{\infty} b_j \frac{z^j}{j!},$$

where the coefficients b_j are given by (2.101), is the ring isomorphism from the ring M_π of O_π-valued measures on O_π onto the ring FD_π of formal divided power series with coefficients from O_π.

For related subjects in the characteristic zero case see [4, 98].

As an example, consider the Dirac measure $\mu = \delta_\alpha$ supported in $\alpha \in O_\pi$.

Proposition 2.34 *The formal divided power series corresponding to the measure δ_α is*

$$P_{\delta_\alpha}(z) = \sum_{j=0}^{\infty} G_j(\alpha)\frac{z^j}{j!}. \qquad (2.104)$$

Proof. By the definition of δ_α,

$$\int_{O_\pi} G_j(t)\, d\delta_\alpha(t) = G_j(\alpha).$$

Now the equality (2.103) is a consequence of (2.101) and (2.102). ∎

The next nice property follows directly from the above definitions and the identity (1.43).

Proposition 2.35 *For any $\alpha, \beta \in O_\pi$,*

$$P_{\delta_{\alpha+\beta}}(z) = P_{\delta_\alpha}(z) P_{\delta_\beta}(z).$$

2.6.3. The ring of hyperderivatives. Another interpretation of the above objects is given in terms of formal differential operators $\sum_{k=0}^{\infty} a_k \dfrac{D^k}{k!}$ where $a_k \in O_\pi$, $D = \dfrac{d}{dt}$. Just as formal divided power series, these objects, called *hyper derivatives*, do not make literal sense. Nevertheless, these formal objects form a ring $Diff_\pi$, and, for example, their action on polynomials is well defined. On the other hand, there is an isomorphism $FD_\pi \to Diff_\pi$ given by

$$P(z) = \sum a_k \frac{z^k}{k!} \mapsto P(D) = \sum a_k \frac{D^k}{k!}.$$

Note that the ring $Diff_\pi$ is not generated by D.

2.7 Comments

Theorems 2.1 and 2.5 were proved by the author in [62]. Note that the case $k = 0$ (the differentiability) of Theorem 2.5 was established much earlier by Wagner [119], and our proof for the general case is based on Wagner's result (see Lemma 2.2). The results about the Volkenborn-type integration (Sections 2.1.3 and 2.1.4) are also taken from [62].

The construction of the fractional derivatives (Section 2.1.5) was given in [67]. Proposition 2.8 was proved earlier by Jeong [55]; our elementary proof follows [67].

The version of umbral calculus expounded in Section 2.2 was developed in [65].

The characterization of locally analytic functions (Section 2.3) was given by Yang [124]; the case of functions analytic on O was considered also in the paper [62], from which the elementary proof of Corollary 2.23 is taken.

A characterization of general C^n-functions (in the sense of Schikhof [98]) in terms of their Carlitz expansions (Theorem 2.26) is a special case of the results by Yang [125] who considered not only the field K, but all finite places of $\mathbb{F}_q(x)$. Wagner [120, 121] studied the conditions for differentiability of a continuous function at a given point from O (in terms of the Carlitz expansion); he also proved Lemma 2.24.

General results about entire functions over K_∞ are taken from [21]; see also [43, 126]. The analog of Pólya's theorem was proved by Adam [1]; for earlier results in this direction see [21, 31].

Our exposition of O_π-valued measures and divided power series follows Goss [43, 44].

3
Differential equations

In this chapter we consider various classes of $\mathbb{F}_q$-linear differential equations with the Carlitz derivatives. We prove existence and uniqueness theorems for regular systems of such equations (the counterparts of regular systems of linear differential equations over $\mathbb{C}$), as well as for strongly nonlinear equations containing self-compositions $y \circ y \circ \cdots \circ y$ of an unknown function y. We consider the behavior of solutions of singular equations; in particular, we introduce and investigate a kind of regular singularity. As a first step in developing a theory of partial differential equations with the Carlitz derivatives, we consider some analogs of classical evolution equations.

3.1 Existence and uniqueness theorems

3.1.1. The framework. As we saw in Section 1.6, the Carlitz exponential e_C satisfies the simplest equation $de_C = e_c$ containing the Carlitz derivative d defined in (1.65). In Chapter 4 we will consider many other special functions satisfying differential equations of this kind. Here our task is to develop a general theory of such equations.

In fact, we consider equations (or systems) with holomorphic or polynomial coefficients. The meaning of a polynomial (holomorphic) coefficient in the function field case is not a usual multiplication by a polynomial (holomorphic function), but the action of a polynomial (holomorphic function) in the operator τ, $\tau u = u^q$. As in classical theory, we have to make a distinction between the regular and singular cases.

In the analytic theory of linear differential equations over $\mathbb{C}$ a regular equation has a constant leading coefficient (which can be assumed equal to 1). A leading coefficient of a singular equation is a holomorphic function having zeros at some points. One can divide the equation by its leading coefficient, but then poles would appear at other coefficients, and the solution can have singularities (not only poles but in general also essential singularities) at those points.

Similarly, in our case we understand a regular equation as one with the coefficient 1 at the highest order derivative. As usual, a regular higher-order equation can be transformed into a regular first-order system. For the regular case we obtain a local existence and uniqueness theorem, which is similar to analogous results for equations over $\mathbb{C}$ or $\mathbb{Q}_p$ (for the latter see [78]). The only difference is a formulation of the initial condition, which is specific for the function field case.

The leading coefficient $A_m(\tau)$ of a singular $\mathbb{F}_q$-linear equation of an order m is a nonconstant holomorhic function of the operator τ. Now one cannot divide the equation

$$A_m(\tau)d^m u(t) + A_{m-1}(\tau)d^{m-1}u(t) + \ldots + A_0(\tau)u(t) = f(t)$$

for an $\mathbb{F}_q$-linear function $u(t)$ (note that automatically $u(0) = 0$) by $A_m(\tau)$. If $A_m(\tau) = \sum\limits_{i=0}^{\infty} a_{mi}\tau^i$, $a_{mi} \in \overline{K}_c$, then

$$A_m(\tau)d^m u = \sum_{i=0}^{\infty} a_{mi} (d^m u)^{q^i},$$

and even when A_m is a polynomial, in order to resolve our equation with respect to $d^m u$ one has to solve an algebraic equation.

Thus for the singular case the situation looks even more complicated than in the classical theory. However we show that the behavior of the solutions cannot be too intricate. Namely, in striking contrast to the classical theory, any formal series solution converges in some (sufficiently small) neighbourhood of the singular point $t = 0$. Note that in the p-adic case a similar phenomenon takes place for equations satisfying certain strong conditions upon zeros of indicial polynomials [26, 102, 8, 86]. In our case such behavior is proved for any equation, which resembles the (much simpler) case [86] of differential equations over a field of characteristics zero, whose residue field also has characteristic zero. Note however that in general a singular equation need not possess a formal power series solution.

3.1.2. Equations without singularities. Let us consider an equation

$$dy(t) = P(\tau)y(t) + f(t) \tag{3.1}$$

where for each $z \in \left(\overline{K}_c\right)^m$, $t \in K$,

$$P(\tau)z = \sum_{k=0}^{\infty} \pi_k z^{q^k}, \quad f(t) = \sum_{j=0}^{\infty} \varphi_j \frac{t^{q^j}}{D_j}, \qquad (3.2)$$

π_k are $m \times m$ matrices with elements from $\overline{K}_c$, $\varphi_j \in \left(\overline{K}_c\right)^m$, and it is assumed that the series (3.2) have positive radii of convergence. The action of the operator τ upon a vector or a matrix is defined componentwise, so that $z^{q^k} = \left(z_1^{q^k}, \ldots, z_m^{q^k}\right)$ for $z = (z_1, \ldots, z_m)$. Similarly, if $\pi = (\pi_{ij})$ is a matrix, we write $\pi^{q^k} = \left(\pi_{ij}^{q^k}\right)$. The norm $|\pi|$ of a matrix π with elements from $\overline{K}_c$ is defined as the maximum of the absolute values of the elements.

We will seek an $\mathbb{F}_q$-linear solution of the (3.1) in some neighborhood of the origin, of the form

$$y(t) = \sum_{i=0}^{\infty} y_i \frac{t^{q^i}}{D_i}, \quad y_i \in \left(\overline{K}_c\right)^m, \qquad (3.3)$$

where y_0 is a given element, so that the "initial" condition for our situation is

$$\lim_{t \to 0} t^{-1} y(t) = y_0. \qquad (3.4)$$

Note that the function (3.3), provided the series has a positive radius of convergence, tends to zero for $t \to 0$, so that the right-hand side of (3.1) makes sense for small $|t|$.

Theorem 3.1 *For any $y_0 \in \left(\overline{K}_c\right)^m$ the equation (3.1) has a unique local solution of the form (3.3), which satisfies (3.4), with the series having a positive radius of convergence.*

Proof. Making (if necessary) the substitutions $t = c_1 t'$, $y = c_2 y'$, with sufficiently small $|c_1|, |c_2|$, we may assume that the coefficients in (3.2) are such that $\varphi_j \to 0$ for $j \to \infty$,

$$|\pi_k^q| \cdot q^{-\frac{q^{k+1}-q}{q-1}} \leq 1, \quad k = 0, 1, \ldots \qquad (3.5)$$

Using the identities

$$d\left(\frac{t^{q^i}}{D_i}\right) = \frac{t^{q^{i-1}}}{D_{i-1}}, \ i \geq 1; \quad d(\text{const}) = 0; \qquad (3.6)$$

$$\tau\left(\frac{t^{q^{i-1}}}{D_{i-1}}\right) = [i]\frac{t^{q^i}}{D_i}, \quad i \geq 1, \tag{3.7}$$

we substitute (3.3) into (3.1), which results in the recurrence formula for the coefficients y_i:

$$y_{l+1} = \sum_{n+k=l} \pi_k^q y_n^{q^{k+1}} [n+1]^{q^k} \ldots [n+k]^q + \varphi_l^q, \quad l = 0, 1, 2, \ldots, \tag{3.8}$$

where the expressions in square brackets are omitted if $k = 0$.

It is seen from (3.8) that a solution of (3.1), (3.4) (if it exists) is unique. Since $|[n]| = q^{-1}$ for all $n > 0$, we find that

$$\left|[n+1]^{q^k} \ldots [n+k]^q\right| = q^{-(q^k + \cdots + q)} = q^{-\frac{q^{k+1}-q}{q-1}},$$

and it follows from (3.5),(3.8) that

$$|y_{l+1}| \leq \max\left\{|\varphi_l|^q, |y_0|^{q^{l+1}}, |y_1|^{q^l}, \ldots, |y_l|^q\right\}.$$

Since $\varphi_n \to 0$, there exists a number l_0 such that $|\varphi_l| \leq 1$ for $l \geq l_0$. Now either $|y_l| \leq 1$ for all $l \geq l_0$ (and then the series (3.3) is convergent in a neighborhood of the origin), or $|y_{l_1}| > 1$ for some $l_1 \geq l_0$. In the latter case

$$|y_{l+1}| \leq \max\left\{|y_0|^{q^{l+1}}, |y_1|^{q^l}, \ldots, |y_l|^q\right\}, \quad l \geq l_1.$$

Let us choose $A > 0$ in such a way that

$$|y_l| \leq A^{q^l}, \quad l = 1, 2, \ldots, l_1.$$

Then it follows easily by induction that $|y_l| \leq A^{q^l}$ for all l, which implies the convergence of (3.3) near the origin. ∎

3.1.3. Singular equations. We will consider scalar equations of arbitrary order

$$\sum_{j=0}^{m} A_j(\tau) d^j u = f \tag{3.9}$$

where

$$f(t) = \sum_{n=0}^{\infty} \varphi_n \frac{t^{q^n}}{D_n},$$

$A_j(\tau)$ are power series having (as well as that for f) positive radii of convergence.

It will be convenient to start from the model equation

$$\sum_{j=0}^{m} a_j \tau^j d^j u = f, \quad a_j \in \overline{K}_c, \ a_m \neq 0. \qquad (3.10)$$

Suppose that $u(t)$ is a formal solution of (3.10), of the form

$$u(t) = \sum_{n=0}^{\infty} u_n \frac{t^{q^n}}{D_n}. \qquad (3.11)$$

Then

$$a_0 \sum_{n=0}^{\infty} u_n \frac{t^{q^n}}{D_n} + \sum_{j=1}^{m} a_j \sum_{n=j}^{\infty} u_n [n-j+1]\ldots[n] \frac{t^{q^n}}{D_n} = \sum_{n=0}^{\infty} \varphi_n \frac{t^{q^n}}{D_n}.$$

Changing the order of summation we find that for $n \geq m$

$$u_n \left(a_0 + \sum_{j=1}^{m} a_j [n-j+1]\ldots[n] \right) = \varphi_n. \qquad (3.12)$$

Let us consider the expression

$$\Phi_n = a_0 + \sum_{j=1}^{m} a_j [n-j+1]\ldots[n], \quad n \geq m.$$

Using repeatedly the identity $[i]^q + [1] = [i+1]$ we find that

$$\Phi_n^{q^m} = a_o^{q^m} + \sum_{j=1}^{m} a_j^{q^m} \prod_{k=0}^{j-1} [n-k]^{q^m}$$

$$= a_o^{q^m} + \sum_{j=1}^{m} a_j^{q^m} \prod_{k=0}^{j-1} \left([n]^{q^{m-k}} - \sum_{l=1}^{k} [1]^{q^{m-l}} \right),$$

that is $\Phi_n^{q^m} = \Phi^{(m)}([n])$ where

$$\Phi^{(m)}(t) = a_o^{q^m} + \sum_{j=1}^{m} a_j^{q^m} \prod_{k=0}^{j-1} \left(t^{q^{m-k}} - \sum_{l=1}^{k} [1]^{q^{m-l}} \right)$$

is a polynomial on $\overline{K}_c$ of a certain degree N not depending on n. Let $\theta_1, \ldots, \theta_N$ be its roots. Then

$$\Phi^{(m)}([n]) = a_m^{q^m} \prod_{\nu=1}^{N} ([n] - \theta_\nu).$$

As $n \to \infty$, $[n] \to -x$ in $\overline{K}_c$. We may assume that $\theta_\nu \neq [n]$ for all ν, if n is large enough. If $\theta_\nu \neq -x$ for all ν, then for large n, say $n \geq n_0 \geq m$,

$$\left|\Phi^{(m)}([n])\right| \geq \mu > 0.$$

If $k \leq N$ roots θ_ν coincide with $-x$, then

$$\left|\Phi^{(m)}([n])\right| \geq \mu q^{-kq^n}, \quad n \geq n_0.$$

Combining the inequalities and taking the root we get

$$|\Phi_n| \geq \mu_1 q^{-\mu_2 q^n}, \quad n \geq n_0. \tag{3.13}$$

where $\mu_1, \mu_2 > 0$. Now it follows from (3.12) and (3.13) that the series (3.11) has (together with the series for f) a positive radius of convergence.

Turning to the general equation (3.9) we note first of all that one can apply an operator series $A(\tau) = \sum_{k=0}^{\infty} \alpha_k \tau^k$ (even without assuming its convergence) to a formal series (3.11), setting

$$\tau^k u(t) = \sum_{n=0}^{\infty} u_n^{q^k} [n+1]^{q^{k-1}} \ldots [n+k] \frac{t^{q^{n+k}}}{D_{n+k}}, \quad k \geq 1,$$

and

$$A(\tau)u(t) = \sum_{l=0}^{\infty} \frac{t^{q^l}}{D_l} \sum_{n+k=l} \alpha_k u_n^{q^k} [n+1]^{q^{k-1}} \ldots [n+k]$$

where the factor $[n+1]^{q^{k-1}} \ldots [n+k]$ is omitted for $k=0$.

Therefore the notion of a formal solution (3.11) makes sense for equation (3.9).

We will need the following elementary estimate.

Lemma 3.2 *Let $k \geq 2$ be a natural number, with a given partition $k = i_1 + \cdots + i_r$, where $i_1, \ldots, i_r$ are positive integers, $r \geq 1$. Then*

$$q^{i_1 + \cdots + i_r} + q^{i_2 + \cdots + i_r} + \cdots + q^{i_r} \leq q^{k+1}.$$

Proof. The assertion is obvious for $k=2$. Suppose it has been proved for some k and consider a partition

$$k + 1 = i_1 + \cdots + i_r.$$

If $i_1 > 1$ then $k = (i_1 - 1) + i_2 + \cdots + i_r$, so that
$$q^{(i_1-1)+i_2+\cdots+i_r} + q^{i_2+\cdots+i_r} + \cdots + q^{i_r} \leq q^{k+1}$$
whence
$$q^{i_1+i_2+\cdots+i_r} + q^{i_2+\cdots+i_r} + \cdots + q^{i_r} \leq q^{k+2}.$$

If $i_1 = 1$ then $k = i_2 + \cdots + i_r$,
$$q^{i_2+\cdots+i_r} + q^{i_3+\cdots+i_r} + \cdots + q^{i_r} \leq q^{k+1}$$
and
$$q^{i_1+\cdots+i_r} + q^{i_2+\cdots+i_r} + \cdots + q^{i_r} \leq 2q^{k+1} \leq q^{k+2}. \blacksquare$$

Now we are ready to formulate the main result of this section.

Theorem 3.3 *Let $u(t)$ be a formal solution (3.11) of equation (3.9), where the series for $A_j(\tau)z$, $z \in \overline{K}_c$, and $f(t)$, have positive radii of convergence. Then the series (3.11) has a positive radius of convergence.*

Proof. Applying (if necessary) the operator τ a sufficient number of times to both sides of (3.9) we may assume that
$$A_j(\tau) = \sum_{i=0}^{\infty} a_{ji}\tau^{i+j}, \quad a_{ji} \in \overline{K}_c, \ j = 0, 1, \ldots, m,$$
where $a_{j0} \neq 0$ at least for one value of j. Let us assume, for example, that $a_{m0} \neq 0$ (otherwise the reasoning below would need an obvious adjustment). Denote by L the operator at the left-hand side of (3.9), and by L_0 its "principal part",
$$L_0 u = \sum_{j=0}^{m} a_{j0}\tau^j d^j u$$
(the model operator considered above; we will maintain the notation introduced there). Note that L_0 is a linear operator.

As we have seen,
$$L_0\left(\frac{t^{q^n}}{D_n}\right) = \Phi_n \frac{t^{q^n}}{D_n}, \quad n \geq n_0,$$

where Φ_n satisfies the inequality (3.13). This means that L_0 is an automorphism of the vector space X of formal series

$$u = \sum_{n=n_0}^{\infty} u_n \frac{t^{q^n}}{D_n}, \quad u_n \in \overline{K}_c,$$

as well as of its subspace Y consisting of series with positive radii of convergence.

Let us write the formal solution u of equation (3.9) as $u = v + w$, where

$$v = \sum_{n=0}^{n_0-1} u_n \frac{t^{q^n}}{D_n}, \quad w = \sum_{n=n_0}^{\infty} u_n \frac{t^{q^n}}{D_n}.$$

Then (3.9) takes the form

$$Lw = g, \tag{3.14}$$

with $g = \sum g_n \frac{t^{q^n}}{D_n} \in Y$. In order to prove our theorem, it is sufficient to verify that $w \in Y$.

For any $y \in X$ we can write

$$Ly = (L_0 - L_1)y = L_0(I - L_0^{-1}L_1)y$$

where

$$L_1 y = -\sum_{j=0}^{m}\sum_{i=1}^{\infty} a_{ji}\tau^{i+j}d^j y. \tag{3.15}$$

In particular, it is seen from (3.14) that

$$(I - L_0^{-1}L_1)w = L_0^{-1}g, \quad L_0^{-1}g \in Y.$$

Writing formally

$$(I - L_0^{-1}L_1)^{-1} = \sum_{k=0}^{\infty} \left(L_0^{-1}L_1\right)^k$$

and noticing that $L_0^{-1}L_1 : X \to \tau X$, we find that

$$w = \sum_{k=0}^{\infty} \left(L_0^{-1}L_1\right)^k h, \tag{3.16}$$

where $h = L_0^{-1}g = \sum_{n=n_0}^{\infty} h_n \frac{t^{q^n}}{D_n}$, $h_n = \Phi_n^{-1} g_n$, and the series in (3.16) converges in the natural non-Archimedean topology of the space X.

Differential equations

A direct calculation shows that for any $\lambda \in \overline{K}_c$

$$(L_0^{-1}L_1)\left(\lambda \frac{t^{q^n}}{D_n}\right) = -\sum_{i=1}^{\infty} \lambda^{q^i} \Phi_{n+i}^{-1} \Psi_i^{(n)} \frac{t^{q^{n+i}}}{D_{n+i}}$$

where

$$\Psi_i^{(n)} = [n+1]^{q^{i-1}}[n+2]^{q^{i-2}} \ldots [n+i] \sum_{j=0}^{m} [n-j+1]^{q^i} \ldots [n]^{q^i} a_{ji},$$

and the coefficient at a_{j0} in the last sum is assumed to equal 1.

Proceeding by induction we get

$$(L_0^{-1}L_1)^r \left(\lambda \frac{t^{q^n}}{D_n}\right)$$
$$= (-1)^r \sum_{i_1,\ldots,i_r=1}^{\infty} \left(\Psi_{i_1}^{(n)}\right)^{q^{i_2+\cdots+i_r}} \left(\Psi_{i_2}^{(n+i_1)}\right)^{q^{i_3+\cdots+i_r}} \cdots$$
$$\times \left(\Psi_{i_r}^{(n+i_1+\cdots+i_{r-1})}\right) \lambda^{q^{i_1+\cdots+i_r}} \Phi_{n+i_1}^{-q^{i_2+\cdots+i_r}} \Phi_{n+i_1+i_2}^{-q^{i_3+\cdots+i_r}} \cdots \Phi_{n+i_1+\cdots+i_r}^{-1}$$
$$\times \frac{t^{q^{n+i_1+\cdots+i_r}}}{D_{n+i_1+\cdots+i_r}}, \quad r = 1,2,\ldots$$

Substituting this into (3.16) and changing the order of summation we find an explicit formula for $w(t)$:

$$w(t)$$
$$= \sum_{l=n_0}^{\infty} \frac{t^{q^l}}{D_l} \sum_{\substack{n+i_1+\cdots+i_r=l \\ n \geq n_0,\, i_i,\ldots,i_r \geq 1}} (-1)^r h_n^{q^{l-n}} \left(\Psi_{i_1}^{(n)}\right)^{q^{i_2+\cdots+i_r}} \left(\Psi_{i_2}^{(n+i_1)}\right)^{q^{i_3+\cdots+i_r}} \cdots$$
$$\times \left(\Psi_{i_r}^{(n+i_1+\cdots+i_{r-1})}\right) \Phi_{n+i_1}^{-q^{i_2+\cdots+i_r}} \Phi_{n+i_1+i_2}^{-q^{i_3+\cdots+i_r}} \cdots \Phi_{n+i_1+\cdots+i_r}^{-1}. \quad (3.17)$$

Observe that

$$\left|\Psi_i^{(n)}\right| \leq (q^{-1})^{q^{i-1}+q^{i-2}+\cdots+1} \sup_j |a_{ji}|, \quad |g_n| \leq M_1^{q^n}, \quad |a_{ji}| \leq M_2^{q^i},$$

$M_1, M_2 \geq 1$ (due to positivity of the corresponding radii of convergence). We have

$$\left|h_n^{q^{l-n}}\right| \leq |\Phi_n|^{-q^{i_1+\cdots+i_r}} M_1^{q^l},$$

and by Lemma 3.2,

$$|\Phi_n|^{-q^{i_1+\cdots+i_r}}|\Phi_{n+i_1}|^{-q^{i_2+\cdots+i_r}}\cdots|\Phi_{n+i_1+\cdots+i_r}|^{-1}$$
$$\leq \mu_1^{q^{i_1+\cdots+i_r}+q^{i_2+\cdots+i_r}+\cdots+q^{i_r}+1}q^{\mu_2(q^{n+i_1+\cdots+i_r}+q^{n+i_2+\cdots+i_r}+\cdots+q^{n+i_r}+q^n)}$$
$$\leq \mu_1^{q^{l-n+1}+1}q^{\mu_2 q^n(q^{l-n+1}+1)} \leq q^{\mu_3 q^{l+1}}, \quad \mu_3 > 0.$$

Lemma 3.2 also yields

$$\left|\Psi_{i_1}^{(n)}\right|^{q^{i_2+\cdots+i_r}}\left|\Psi_{i_2}^{(n+i_1)}\right|^{q^{i_3+\cdots+i_r}}\cdots\left|\Psi_{i_r}^{(n+i_1+\cdots+i_{r-1})}\right|$$
$$\leq M_2^{q^{i_1+\cdots+i_r}+q^{i_2+\cdots+i_r}+\cdots+q^{i_r}} \leq M_2^{q^{l+1}}.$$

Writing (3.17) as

$$w(t) = \sum_{l=n_0}^{\infty} w_l \frac{t^{q^l}}{D_l}$$

we find that

$$\limsup_{l\to\infty} |w_l|^{q^{-l}} \leq \limsup_{l\to\infty} \left(q^{\mu_3 q^{l+1}} M_1^{q^l} M_2^{q^{l+1}}\right)^{q^{-l}} < \infty,$$

which implies the positivity of the radius of convergence. ∎

3.2 Strongly nonlinear equations

3.2.1. Recurrence relations. Studying strongly nonlinear equations and implicit functions we encounter recurrence relations of the same form

$$c_{i+1} = \mu_i \sum_{\substack{j+l=i \\ l\neq 0}} \sum_{k=1}^{\infty} B_{jkl} \left(\sum_{n_1+\cdots+n_k=l} c_{n_1} c_{n_2}^{q^{n_1}} \cdots c_{n_k}^{q^{n_1+\cdots+n_{k-1}}}\right)^{q^{j+\lambda}} + a_i,$$

$$i = 1, 2, \ldots, \quad (3.18)$$

(here and below $n_1,\ldots,n_k \geq 1$ in the internal sum), with coefficients from $\overline{K}_c$, such that $|\mu_i| \leq M$, $M > 0$, $|B_{jkl}| \leq B^{kq^j}$, $B \geq 1$, $|a_i| \leq M$ for all i,j,k,l; the number λ is either equal to 1, or $\lambda = 0$, and in that case $|B_{01l}| \leq 1$.

Proposition 3.4 *For an arbitrary element $c_1 \in \overline{K}_c$, the sequence determined by the relation (3.18) satisfies the estimate $|c_n| \leq C^{q^n}$, $n = 1, 2, \ldots$, with some constant $C \geq 1$.*

Proof. Set $c_n = \sigma d_n$, $|\sigma| < 1$, $n = 1, 2, \ldots$, and substitute this into (3.18). We have

$$d_{i+1} = \mu_i \left\{ \sum_{\substack{j+l=i \\ l \neq 0}} \sum_{k=1}^{\infty} B_{jkl} \sum_{n_1+\cdots+n_k=l} \sigma^{\left(1+q^{n_1}+\cdots+q^{n_1+\cdots+n_{k-1}}\right)q^{j+\lambda}-1} \right.$$

$$\left. \times \left(d_{n_1} d_{n_2}^{q^{n_1}} \cdots d_{n_k}^{q^{n_1+\cdots+n_{k-1}}} \right)^{q^{j+\lambda}} \right\} + \sigma^{-1} a_i.$$

Here

$$\left| \sigma^{\left(1+q^{n_1}+\cdots+q^{n_1+\cdots+n_{k-1}}\right)q^{j+\lambda}-1} \right| \leq |\sigma|^{kq^{j+\lambda}-1},$$

and (under our assumptions) choosing σ such that $|\sigma|$ is small enough we reduce (3.18) to the relation

$$d_{i+1} = \mu_i \sum_{\substack{j+l=i \\ l \neq 0}} \sum_{k=1}^{\infty} b_{jkl} \sum_{n_1+\cdots+n_k=l} \left(d_{n_1} d_{n_2}^{q^{n_1}} \cdots d_{n_k}^{q^{n_1+\cdots+n_{k-1}}} \right)^{q^{j+\lambda}}$$

$$+ \sigma^{-1} a_i, \quad i = 1, 2, \ldots, \quad (3.19)$$

where $|b_{jkl}| \leq 1$.

It follows from (3.19) that

$$|d_{i+1}| \leq M \max_{\substack{j+l=i \\ l \neq 0}} \sup_{k \geq 1} \max_{n_1+\cdots+n_k=l} \max \left\{ \left(|d_{n_1}| \cdot |d_{n_2}^{q^{n_1}}| \cdots \right.\right.$$

$$\left.\left. \times |d_{n_k}|^{q^{n_1+\cdots+n_{k-1}}} \right)^{q^{j+\lambda}}, M^{-1} |\sigma^{-1} a_i| \right\}.$$

Let $B = \max \left\{ 1, M, |d_1|, M^{-1} \sup_i |\sigma^{-1} a_i| \right\}$. Let us show that

$$|d_n| \leq B^{q^{n-1}+q^{n-2}+\cdots+1}, \quad n = 1, 2, \ldots. \quad (3.20)$$

This is obvious for $n = 1$. Suppose that we have proved (3.20) for $n \leq i$.

Then

$$|d_{i+1}| \leq M \max_{\substack{j+l=i}} \sup_{k \geq 1} \max_{n_1+\cdots+n_k=l} \left(B^{q^{n_1-1}+q^{n_1-2}+\cdots+1} \right.$$
$$\times B^{q^{n_1+n_2-1}+q^{n_1+n_2-2}+\cdots+q^{n_1}} \cdots$$
$$\times \cdots B^{q^{n_1+\cdots+n_{k-1}+n_k-1}+q^{n_1+\cdots+n_{k-1}+n_k-2}+\cdots+1} + q^{n_1+\cdots+n_{k-1}} \right)^{q^{j+1}}$$
$$\leq M \max_{j+l=i} B^{q^{j+l}+\cdots+q^{j+1}} \leq B \cdot B^{q^i+q^{i-1}+\cdots+q} = B^{q^i+q^{i-1}+\cdots+1},$$

and we have proved (3.20). Therefore

$$|c_n| \leq |\sigma| B^{\frac{q^n-1}{q-1}} \leq C^{q^n}$$

for some C, as desired. ∎

3.2.2. Implicit functions of algebraic type. In this section we look for $\mathbb{F}_q$-linear locally holomorphic solutions of equations of the form

$$P_0(t) + P_1(t) \circ z + P_2(t) \circ (z \circ z) + \cdots + P_N(t) \circ \underbrace{(z \circ z \circ \cdots \circ z)}_{N} = 0 \quad (3.21)$$

where $P_0, P_1, \ldots P_N \in \mathcal{R}_{\overline{K}_c}$ (see Section 1.1.2 for the definitions). Suppose that the coefficient $P_k(t) = \sum_{j \geq 0} a_{jk} t^{q^j}$ is such that $a_{00} = 0$, $a_{01} \neq 0$; these assumptions are similar to those guaranteeing the existence and uniqueness of a solution in classical complex analysis. Then (see Section 1.1.2) P_1 is invertible in $\mathcal{R}_{\overline{K}_c}$, and we can rewrite (3.21) in the form

$$z + Q_2(t) \circ (z \circ z) + \cdots + Q_N(t) \circ \underbrace{(z \circ z \circ \cdots \circ z)}_{N} = Q_0(t) \quad (3.22)$$

where $Q_0, Q_2, \ldots, Q_N \in \mathcal{R}_{\overline{K}_c}$, that is

$$Q_k(t) = \sum_{j=0}^{\infty} b_{jk} t^{q^j}, \quad |b_{jk}| \leq B_k^{q^j},$$

for some constants $B_k > 0$, and $b_{00} = 0$.

Proposition 3.5 *Equation (3.21) has a unique solution $z \in \mathcal{R}_{\overline{K}_c}$ satisfying the "initial condition"*

$$\frac{z(t)}{t} \longrightarrow 0, \quad t \to 0.$$

Proof. Let us look for a solution of the transformed equation (3.22), of the form
$$z(t) = \sum_{i=1}^{\infty} c_i t^{q^i}, \quad c_i \in \overline{K}_c; \qquad (3.23)$$
our initial condition is automatically satisfied for a function (3.23).

Substituting (3.23) into (3.22) we come to the system of equalities

$$c_i = -\sum_{k=2}^{N} \sum_{\substack{j+l=i \\ j\geq 0, l\geq 1}} b_{jk} \left(\sum_{\substack{n_1+\cdots+n_k=l \\ n_j \geq 1}} c_{n_1} c_{n_2}^{q^{n_1}} \cdots c_{n_k}^{q^{n_1+\cdots+n_{k-1}}} \right)^{q^j} + b_{i0}, \quad i \geq 1. \qquad (3.24)$$

In each of them the right-hand side depends only on $c_1, \ldots, c_{i-1}$, so that the relations (3.24) determine the coefficients of a solution (3.23) uniquely. By Proposition 3.4, $z \in \mathcal{R}_{\overline{K}_c}$. ∎

More generally, let
$$P_1(t) = \sum_{j \geq \nu} a_{j1} t^{q^j}, \quad \nu \geq 0, \quad a_{\nu 1} \neq 0.$$

Then equation (3.21) has a unique solution in $\mathcal{R}_{\overline{K}_c}$, of the form
$$z(t) = \sum_{i=\nu+1}^{\infty} c_i t^{q^i}, \quad c_i \in \overline{K}_c.$$

The proof is similar.

3.2.3. Existence and uniqueness of solutions. Let us consider the equation
$$dz(t) = \sum_{j=0}^{\infty} \sum_{k=1}^{\infty} a_{jk} \tau^j \underbrace{(z \circ z \circ \cdots \circ z)}_{k}(t) + \sum_{j=0}^{\infty} a_{j0} t^{q^j} \qquad (3.25)$$

where $a_{jk} \in \overline{K}_c$, $|a_{jk}| \leq A^{kq^j}$ ($k \geq 1$), $|a_{j0}| \leq A^{q^j}$, $A \geq 1$. We look for a solution in the class of $\mathbb{F}_q$-linear locally holomorphic functions of the form
$$z(t) = \sum_{k=1}^{\infty} c_k t^{q^k}, \quad c_k \in \overline{K}_c, \qquad (3.26)$$

thus assuming the initial condition $t^{-1} z(t) \to 0$, as $t \to 0$.

Theorem 3.6 *A solution (3.26) of equation (3.25) exists with a non-zero radius of convergence, and is unique.*

Proof. We may assume that

$$|a_{j0}| \leq 1, \quad a_{j0} \to 0, \quad \text{as } j \to \infty. \tag{3.27}$$

Indeed, if that is not satisfied, we can perform a time change $t = \gamma t_1$ obtaining an equation of the same form, but with the coefficients $a_{j0}\gamma^{q^j}$ instead of a_{j0}, and it remains to choose γ with $|\gamma|$ small enough. Note that, in contrast to the case of the usual derivatives, the operator d commutes with the above time change.

Assuming (3.27) we substitute (3.26) into (3.25) using the fact that $d\left(c_k t^{q^k}\right) = c_k^{1/q}[k]^{1/q} t^{q^{k-1}}$, $k \geq 1$, where $[k] = x^{q^k} - x$. Comparing the coefficients we come to the recursion

$$c_{i+1} = [i+1]^{-1} \sum_{\substack{j+l=i \\ j\geq 0, l\geq 1}} \sum_{k=1}^{\infty} a_{jk}^q \left(\sum_{n_1+\cdots+n_k=l} c_{n_1} c_{n_2}^{q^{n_1}} \cdots c_{n_k}^{q^{n_1+\cdots+n_{k-1}}} \right)^{q^{j+1}}$$
$$+ a_{i0}, \quad i \geq 1,$$

where $c_1 = [1]^{-1} a_{00}^q$. This already shows the uniqueness of a solution. The fact that $|c_i| \leq C^{q^i}$ for some C follows from Proposition 3.4. ∎

Using Proposition 3.5 we can easily reduce to the form (3.25) some classes of equations given in the form not resolved with respect to dz.

As in the classical case of equations over $\mathbb{C}$ (see [51]), some of equations (3.25) can have also nonholomorphic solutions, in particular those which are meromorphic in the sense of Section 1.1.2. As an example, we consider Riccati-type equations

$$dy(t) = \lambda(y \circ y)(t) + (P(\tau)y)(t) + R(t) \tag{3.28}$$

where $\lambda \in \overline{K}_c$, $0 < |\lambda| \leq q^{-1/q^2}$,

$$(P(\tau)y)(t) = \sum_{k=1}^{\infty} p_k y^{q^k}(t), \quad R(t) = \sum_{k=0}^{\infty} r_k t^{q^k},$$

$p_k, r_k \in \overline{K}_c$, $|p_k| \leq q^{-1/q^2}$, $|r_k| \leq q^{-1/q^2}$ for all k.

Differential equations 119

Theorem 3.7 *Under the above assumptions, equation (3.28) possesses solutions of the form*

$$y(t) = ct^{1/q} + \sum_{n=0}^{\infty} a_n t^{q^n}, \quad c, a_n \in \overline{K}_c, \ c \neq 0, \tag{3.29}$$

where the series converges on the open unit disk $|t| < 1$.

Proof. For the function (3.29) we have

$$dy(t) = c^{1/q}[-1]^{1/q} t^{q^{-2}} + \sum_{n=1}^{\infty} a_n^{1/q}[n]^{1/q} t^{q^{n-1}}, \quad [-1] = x^{1/q} - x,$$

$$(y \circ y)(t) = c \left(ct^{1/q} + \sum_{n=0}^{\infty} a_n t^{q^n} \right)^{1/q} + \sum_{n=0}^{\infty} a_n \left(ct^{1/q} + \sum_{m=0}^{\infty} a_m t^{q^m} \right)^{q^n}$$

$$= c^{1+\frac{1}{q}} t^{q^{-2}} + \left(ca_0^{1/q} + ca_0 \right) t^{q^{-1}} + \sum_{n=0}^{\infty} \left(ca_{n+1}^{1/q} + c^{q^{n+1}} a_{n+1} \right) t^{q^n}$$

$$+ \sum_{l=0}^{\infty} t^{q^l} \sum_{\substack{m+n=l \\ m,n \geq 0}} a_n a_m^{q^n}.$$

Finally,

$$(P(\tau)y)(t) = \sum_{k=0}^{\infty} p_{k+1} c^{q^{k+1}} t^{q^k} + \sum_{l=0}^{\infty} t^{q^l} \sum_{\substack{i+j=l \\ i \geq 1, j \geq 0}} p_i a_j^{q^i}.$$

Comparing the coefficients we find that

$$c = \lambda^{-1}[-1]^{1/q}, \quad a_0^{1/q} + a_0 = 0, \tag{3.30}$$

$$a_{l+1}^{1/q}([l+1]^{1/q} - \lambda c) - \lambda c^{q^{l+1}} a_{l+1} = \lambda \sum_{\substack{m+n=l \\ m,n \geq 0}} a_n a_m^{q^n} + \sum_{\substack{i+j=l \\ i \geq 1, j \geq 0}} p_i a_j^{q^i} + r_l, \quad l \geq 0. \tag{3.31}$$

By (3.30), we have $|c| \geq 1$, and either $a_0 = 0$, or $|a_0| = 1$. Next, (3.31) is a recurrence relation (with an algebraic equation to be solved at each step) giving values of a_l for all $l \geq 1$. Let us prove that $|a_j| \leq 1$ for all j. Suppose we have proved that for $j \leq l$. It follows from (3.31) that

$$\left| a_{l+1}[l+1] - \lambda^q c^q a_{l+1} - \lambda^q c^{q^{l+2}} a_{l+1}^q \right| \leq q^{-1/q}. \tag{3.32}$$

120   *Chapter 3*

Suppose that $|a_{l+1}| > 1$. We have $\lambda^q c^q = [-1]$, so that $|\lambda^q c^q| = q^{-1/q}$, and since $|[l+1]| = q^{-1}$ and $|c| \geq 1$, we find that

$$|a_{l+1}[l+1]| < |\lambda^q c^q a_{l+1}| < \left|\lambda^q c^{q^{l+2}} a_{l+1}^q\right|.$$

Therefore the left-hand side of (3.32) equals $|\lambda^q c^q| \cdot \left|c^{q^{l+1}}\right| \cdot |a_{l+1}^q| > q^{-1/q}$, and we have come to a contradiction. ∎

3.3 Regular singularity

3.3.1. Background. Let us consider an analog of the class of equations with regular singularity, the most thoroughly studied class of singular equations (see [27] or [47] for the classical theory of differential equations over $\mathbb{C}$; the case of non-Archimedean fields of characteristic zero was studied in [35, 89]).

A typical class of systems with regular singularity at the origin $\zeta = 0$ over $\mathbb{C}$ consists of systems of the form

$$\zeta y'(\zeta) = \left(B + \sum_{k=1}^{\infty} A_k \zeta^k\right) y(\zeta)$$

where B, A_j are constant matrices, and the series converges on a neighborhood of the origin. Such a system possesses a fundamental matrix solution of the form $W(\zeta)\zeta^C$ where $W(\zeta)$ is holomorphic on a neighborhood of zero, C is a constant matrix, and $\zeta^C = \exp(C \log \zeta)$ is defined by the obvious power series. Under some additional assumptions regarding the eigenvalues of the matrix B, one can take $C = B$. For similar results over $\mathbb{C}_p$ see Section III.8 in [35].

In order to investigate such a class of equations in the framework of $\mathbb{F}_q$-linear analysis over K, one has to go beyond the class of functions represented by power series. An analog of the power function need not be holomorphic, and cannot be defined as above. Fortunately, we have another option here – instead of power series expansions we can use the expansions in Carlitz polynomials on the compact ring of integers $O \subset K$. It is important to stress that our approach would fail if we consider equations over $\overline{K}_c$ instead of K (our solutions may take their values from $\overline{K}_c$, but they are defined over subsets of K). In this sense our techniques are different from those developed for both the characteristic zero cases.

3.3.2. A model scalar equation. We begin with the simplest scalar

Differential equations

equation
$$\tau du = \lambda u, \quad \lambda \in \overline{K}_c, \tag{3.33}$$
whose solution may be seen as a function field counterpart of the power function $t \mapsto t^\lambda$. We look for a continuous $\mathbb{F}_q$-linear solution
$$u(t) = \sum_{i=0}^{\infty} c_i f_i(t), \quad t \in O, \tag{3.34}$$
where $\overline{K}_c \ni c_i \to 0$, as $i \to \infty$, $\{f_i\}$ is the sequence of orthonormal Carlitz polynomials.

It follows from the identities (1.12) and (1.21) that
$$\Delta f_i = [i] f_i + f_{i-1}, \quad i \geq 1; \tag{3.35}$$
as we know, $\Delta f_0 = 0$. Therefore
$$\Delta u(t) = \sum_{j=0}^{\infty} (c_{j+1} + [j] c_j) f_j(t).$$
Substituting into (3.33) and using uniqueness of the Carlitz expansion we find a recurrence relation
$$c_{j+1} + [j] c_j = \lambda c_j, \quad j = 0, 1, 2, \ldots,$$
whence, given c_0, the solution is determined uniquely by
$$c_n = c_0 \prod_{j=0}^{n-1} (\lambda - [j]).$$

Suppose that $|\lambda| \geq 1$. Since $|[j]| = q^{-1}$ for $j \geq 1$, we see that $|c_n| = |c_0| \cdot |\lambda|^n \not\to 0$ if $c_0 \neq 0$. This contradiction shows that equation (3.33) has no continuous solutions if $|\lambda| \geq 1$. Therefore we shall assume that $|\lambda| < 1$. Let $u(t, \lambda)$ be the solution of (3.33) with $c_0 = 1$; note that the fixation of c_0 is equivalent to the initial condition $u(1, \lambda) = c_0$. The function $u(t, \lambda)$ is a function field counterpart of the power function t^λ.

Theorem 3.8 *The function $t \mapsto u(t, \lambda)$, $|\lambda| < 1$, is continuous on O. It is analytic on O if and only if $\lambda = [j]$ for some $j \geq 0$; in this case $u(t, \lambda) = u(t, [j]) = t^{q^j}$. If $\lambda \neq [j]$ for any integer $j \geq 0$, then $u(t, \lambda)$ is locally analytic on O if and only if $\lambda = -x$, and in that case $u(t, -x) = 0$ for $|t| \leq q^{-1}$. The relation*
$$u(t^{q^m}, \lambda) = u(t, \lambda^{q^m} + [m]), \quad t \in O, \tag{3.36}$$

holds for all λ, $|\lambda| < 1$, and for all $m = 0, 1, 2, \ldots$

Proof. If $u(t) = t^{q^j}$, $j \geq 0$, then

$$\Delta u(t) = (xt)^{q^j} - xt^{q^j} = \left(x^{q^j} - x\right) t^{q^j} = [j]u(t),$$

so that $u(t, [j]) = t^{q^j}$.

Suppose that $\lambda \neq [j]$, $j = 0, 1, 2, \ldots$ Then $|c_n| \leq \{\max(|\lambda|, q^{-1})\}^n \to 0$ as $n \to \infty$, so that $u(t, \lambda)$ is continuous. More precisely, if $\lambda \neq -x$, then $|\lambda + x| = q^{-\nu}$ for some $\nu > 0$,

$$|\lambda - [j]| = \left|(\lambda + x) - x^{q^j}\right| = q^{-\nu}, \quad j \geq j_0,$$

if j_0 is large enough. This means that for some positive constant C

$$|c_n| = Cq^{-n\nu}, \quad n \geq j_0. \tag{3.37}$$

On the other hand, if $\lambda = -x$, then $|\lambda - [j]| = q^{-q^j}$,

$$|c_n| = q^{-\frac{q^n - 1}{q - 1}}. \tag{3.38}$$

By Corollary 2.22, the function u is locally analytic if and only if

$$\gamma = \liminf_{n \to \infty} \left\{-q^n \log_q |c_n|\right\} > 0. \tag{3.39}$$

If $\lambda \neq -x$, then by (3.37) $\gamma = 0$, so that $u(t, \lambda)$ is not locally analytic. If $\lambda = -x$, we see from (3.39) and (3.38) that $\gamma = (q-1)^{-1}$, $l = 1$, and $u(t, -x)$ is analytic on any ball of radius q^{-1}. We have

$$u(t, -x) = \sum_{n=0}^{\infty} (-1)^n x^{\frac{q^n - 1}{q - 1}} f_n(t),$$

and $u(t, -x)$ is not identically zero on O due to the uniqueness of the Fourier–Carlitz expansion.

At the same time, since $u(t, -x)$ is analytic on the ball $\{|t| \leq q^{-1}\}$, we can write

$$u(t, -x) = \sum_{m=0}^{\infty} a_m t^{q^m}, \quad |t| \leq q^{-1}.$$

Substituting this into equation (3.33) with $\lambda = -x$, we find that $a_m = 0$ for all m, that is $u(t, -x) = 0$ for $|t| \leq q^{-1}$.

In order to prove (3.36), note first that (3.36) holds for $\lambda = [j]$, $j = 0, 1, 2, \ldots$ Indeed,
$$u(t^{q^m}, [j]) = \left(t^{q^m}\right)^{q^j} = t^{q^{m+j}}$$
and
$$[j]^{q^m} + [m] = \left(x^{q^j} - x\right)^{q^m} + x^{q^m} - x = [m+j].$$
Let us fix $t \in O$. We have
$$u(t, \lambda) = \sum_{n=0}^{\infty} \left\{ \prod_{j=0}^{n-1} (\lambda - [j]) \right\} f_n(t),$$
and the series converges uniformly with respect to $\lambda \in \overline{P}_r$ where
$$\overline{P}_r = \{\lambda \in \overline{K}_c : |\lambda| \leq r\},$$
for any positive $r < 1$. Thus $u(t, \lambda)$ is an analytic element on $\overline{P}_r$ (see Chapter 10 of [36]). Similarly, $u(t^{q^m}, \lambda)$ and $u(t, \lambda^{q^m} + [m])$ are analytic elements on $\overline{P}_r$ (for the latter see Theorem 11.2 from [36]). Suppose that $q^{-1} \leq r < 1$. Since both sides of (3.36) coincide on an infinite sequence of points $\lambda = [j]$, $j = 0, 1, 2, \ldots$, they coincide on $\overline{P}_r$ (see Corollary 23.8 in [36]). This implies their coincidence for $|\lambda| < 1$. ∎

The paradoxical fact that a locally analytic function $u(t, -x)$ vanishes on an open subset is a good illustration of the violation of the principle of analytic continuation in the non-Archimedean case. It is also interesting that the nonlinear equation $du = \lambda u$ is much simpler than the linear equation $\tau du = \lambda u$.

If in (3.33) λ is an $m \times m$ matrix with elements from $\overline{K}_c$ (we shall write $\lambda \in M_m(\overline{K}_c)$), and we look for a solution $u \in M_m(\overline{K}_c)$, then we can find a continuous solution (3.34) with matrix coefficients

$$c_i = \left\{ \prod_{j=0}^{i-1} (\lambda - [j]I_m) \right\} c_0, \quad i \geq 1 \tag{3.40}$$

(I_m is a unit matrix) if $|\lambda| \stackrel{\text{def}}{=} \max |\lambda_{ij}| < 1$. Note that $c_0 = u(1)$, so that if c_0 is an invertible matrix, then u is invertible on a certain neighborhood of 1.

3.3.3. First-order systems. Let us consider a system
$$\tau du - P(\tau)u = 0 \tag{3.41}$$
with the coefficient $P(\tau)$ given in (3.2). We assume that $|\pi_k| \leq \gamma$, $\gamma > 0$, for all k, $|\pi_0| < 1$. Denote by $g(t)$ a solution of the equation $\tau dg = \pi_0 g$. Let $\lambda_1, \ldots, \lambda_m \in \overline{K}_c$ be the eigenvalues of the matrix π_0.

Theorem 3.9 *If*
$$\lambda_i - \lambda_j^{q^k} \neq [k], \quad i,j = 1, \ldots, m; \ k = 1, 2, \ldots, \tag{3.42}$$
then the system (3.41) has a matrix solution
$$u(t) = W(g(t)), \quad W(s) = \sum_{k=0}^{\infty} w_k s^{q^k}, \quad w_0 = I_m, \tag{3.43}$$
where the series for W has a positive radius of convergence.

Proof. Substituting (3.43) into (3.41), using the fact that $\Delta = \tau d$ is a derivation of the composition ring of $\mathbb{F}_q$-linear series, and that $\Delta(t^{q^k}) = [k]t^{q^k}$, we come to the identity
$$\sum_{k=0}^{\infty}[k]w_k \tau^k(g(t)) + \sum_{k=0}^{\infty} w_k \tau^k(\pi_0)\tau^k(g(t)) - \sum_{j=0}^{\infty} \pi_j \tau^j \left(\sum_{k=0}^{\infty} w_k \tau^k(g(t)) \right) = 0. \tag{3.44}$$

If the series for W has indeed a positive radius of convergence (which will be proved later), then all the expressions in (3.44) make sense for small $|t|$, since $g(t) \to 0$ as $|t| \to 0$. Since $w_0 = I_m$, the first summand in the second sum in (3.44) and the summand with $j = k = 0$ in the third sum are cancelled. Changing the order of summation we find that (3.44) is equivalent to the system of equations
$$w_k \left([k]I_m + \tau^k(\pi_0) \right) - \pi_0 w_k = \sum_{j=1}^{k} \pi_j \tau^j (w_{k-j}), \quad k = 1, 2, \ldots, \tag{3.45}$$
with respect to the matrices w_k.

The system (3.45) is solved step by step – if the right-hand side of (3.45) with some k is already known, then w_k is determined uniquely, provided the spectra of the matrices $[k]I_m + \tau^k(\pi_0)$ and π_0 have an empty intersection ([39], Section VIII.1). This condition is equivalent to (3.42), and it remains to prove that the series for W has a nonzero radius of convergence.

Differential equations

Let us transform π_0 to its Jordan normal form. We have $U^{-1}\pi_0 U = A$ where U is an invertible matrix over $\overline{K}_c$, and A is block-diagonal:

$$A = \bigoplus_{\alpha=1}^{l} \left(\lambda^{(\alpha)} I_{d_\alpha} + H^{(\alpha)} \right)$$

where $\lambda^{(\alpha)}$ are eigenvalues from the collection $\{\lambda_1, \ldots, \lambda_m\}$, $H^{(\alpha)}$ is a Jordan cell of order d_α having zeros on the principal diagonal and ones on the one below it. Denote $\mu_k^{(\alpha)} = \left(\lambda^{(\alpha)}\right)^{q^k} + [k]$. If $V_k = \tau^k(U)$, then

$$V_k^{-1}\left([k]I_m + \tau^k(\pi_0)\right)V_k = \bigoplus_{\alpha=1}^{l}\left(\mu_k^{(\alpha)} I_{d_\alpha} + H^{(\alpha)}\right). \qquad (3.46)$$

If B_k is the matrix (3.46), and C_k is the matrix in the right-hand side of (3.45), then (3.45) takes the form

$$w_k V_k B_k V_k^{-1} - UAU^{-1}w_k = C_k$$

or, if we use the notation $\widetilde{w}_k = U^{-1}w_k V_k$,

$$\widetilde{w}_k B_k - A\widetilde{w}_k = \widetilde{C}_k, \quad k = 1, 2, \ldots, \qquad (3.47)$$

where

$$\widetilde{C}_k = U^{-1} C_k V_k = U^{-1}\left(\sum_{j=1}^{k} \pi_j \tau^j \left(U\widetilde{w}_{k-j} V_{k-j}^{-1}\right)\right) V_k$$

$$= U^{-1} \sum_{j=1}^{k} \pi_j \tau^j(U) \tau^j\left(\widetilde{w}_{k-j}\right),$$

$\widetilde{w}_0 = I_m$. We may assume that $|U| \leq 1$, $|U^{-1}| \leq \rho$, $\rho > 0$.

In accordance with the quasidiagonal form of the matrices A and B_k we can decompose the matrix $\widetilde{w}_k$ into $d_\alpha \times d_\beta$ blocks

$$\widetilde{w}_k = \left(\widetilde{w}_k^{(\alpha\beta)}\right), \quad \alpha, \beta = 1, \ldots, l.$$

Similarly we write $\widetilde{C}_k = \left(\widetilde{C}_k^{(\alpha\beta)}\right)$. Then the system (3.47) is decoupled into a system of equations for each block:

$$\left(\mu_k^{(\beta)} - \lambda^{(\alpha)}\right)\widetilde{w}_k^{(\alpha\beta)} - H^{(\alpha)}\widetilde{w}_k^{(\alpha\beta)} + \widetilde{w}_k^{(\alpha\beta)} H^{(\beta)} = \widetilde{C}_k^{(\alpha\beta)}. \qquad (3.48)$$

Equation (3.48) can be considered as a system of scalar equations with respect to elements of the matrix $\widetilde{w}_k^{(\alpha\beta)}$. Let us enumerate these elements

$\left(\widetilde{w}_k^{(\alpha\beta)}\right)_{ij}$ lexicographically (in i,j) with the inverse enumeration order of the second index j. The product $H^{(\alpha)}\widetilde{w}_k^{(\alpha\beta)}$ is obtained from $\widetilde{w}_k^{(\alpha\beta)}$ by the shift of all the rows one step upwards, the last row being filled by zeros. Similarly, the product $\widetilde{w}_k^{(\alpha\beta)}H^{(\beta)}$ is the result of shifting all the columns of $\widetilde{w}_k^{(\alpha\beta)}$ one step to the right and filling the first column by zeros ([39], Section I.3). This means that the system (3.48) (with fixed α,β) with the above enumeration of the unknowns is upper triangular. Indeed, the latter is equivalent to the fact that each equation contains, together with some unknown, only the unknowns with larger numbers, and this property is the result of the above shifts.

Therefore the determinant $D_k^{(\alpha\beta)}$ of the system (3.48) equals $\left(\mu_k^{(\beta)} - \lambda^{(\alpha)}\right)^{d_\alpha d_\beta}$. By our assumption $|\pi_0| < 1$, and if $\lambda^{(\alpha)}$ is an eigenvalue of π_0 with an eigenvector $f \neq 0$, then $|\lambda^{(\alpha)}| \cdot |f| = |\pi_0 f| < |f|$, so that $|\lambda^{(\alpha)}| < 1$. This means that all the coefficients on the left in (3.48) have absolute values ≤ 1.

It follows from (3.42) that $\lambda_i \neq -x$, $i = 1, \ldots, n$. As $k \to \infty$, $\mu_k^{(\beta)} = \left(\lambda^{(\beta)}\right)^{q^k} + x^{q^k} - x \to -x$. Thus $\left|\mu_k^{(\beta)} - \lambda^{(\alpha)}\right| \geq \sigma_1 > 0$ for all k, whence $\left|D_k^{(\alpha\beta)}\right| \geq \sigma_2 > 0$ where σ_2 does not depend on k. Now we obtain an estimate for the solution of the system (3.47),

$$|\widetilde{w}_k| \leq \rho_1 \left|\widetilde{C}_k\right|, \qquad (3.49)$$

with $\rho_1 > 0$ independent of k.

Looking at (3.49) we find that

$$|\widetilde{w}_k| \leq \rho_2 \max_{1 \leq j \leq k} |\widetilde{w}_{k-j}|^{q^j}$$

where ρ_2 does not depend on k. We may assume that $\rho_2 \geq 1$. Now we find that

$$|\widetilde{w}_k| \leq \rho_2^{q^{k-1}+q^{k-2}+\cdots+1}, \quad k = 1, 2, \ldots \qquad (3.50)$$

Indeed, (3.50) is obvious for $k = 1$. Suppose that we have proved the inequalities

$$|\widetilde{w}_j| \leq \rho_2^{q^{j-1}+q^{j-2}+\cdots+1}, \quad 1 \leq j \leq k-1.$$

Then

$$|\widetilde{w}_k| \le \rho_2 \max\left(1, |\widetilde{w}_1|^{q^{k-1}}, \ldots, |\widetilde{w}_{k-1}|^q\right)$$
$$\le \rho_2 \max\left(1, \rho_2^{q^{k-1}}, \rho_2^{(q+1)q^{k-2}}, \ldots, \rho_2^{(q^{k-2}+\cdots+1)q}\right) = \rho_2^{q^{k-1}+q^{k-2}+\cdots+1},$$

and (3.50) is proved.

Therefore, since $w_k = U\widetilde{w}_k \tau^k(U^{-1})$, we have

$$|w_k| \le \rho^{q^k} \cdot \rho_2^{q^{k-1}+\cdots+1} \le \rho_3^{\frac{q^{k+1}-1}{q-1}}, \quad \rho_3 > 0,$$

which means that the series in (3.43) has a positive radius of convergence. ∎

Remarks 3.1 (1) If $\varphi \in \mathbf{F}_q^m$, then, as usual, $v = u\varphi$ is a vector solution of the system $\tau dv - P(\tau)v = 0$, since the system is $\mathbb{F}_q$-linear. However, the system is nonlinear over $\overline{K}_c$, so that we cannot obtain a vector solution in such a way for an arbitrary $\varphi \in \overline{K}_c$.

(2) Analogs of the condition (3.42) occur also in the analytic theory of differential equations over $\mathbb{C}$ (see Corollary 11.2 in [47]) and $\mathbb{Q}_p$ (Section III.8 in [35]). For systems over $\mathbb{C}$, it is requested that differences of the eigenvalues of the leading coefficient π_0 must not be nonzero integers. Over $\mathbb{Q}_p$, in addition to that, the eigenvalues must not be nonzero integers themselves. In both the characteristic zero cases it is possible to get rid of such conditions by using special changes of variables called shearing transformations. For example, let $m = 1$, and the equation over $\mathbb{Q}_p$ has the form

$$\zeta u'(\zeta) = nu(\zeta) + \left(\sum_{k=1}^\infty \pi_k \zeta^k\right) u(\zeta), \quad n \in \mathbb{N}.$$

Then the change of variables $u(\zeta) = \zeta v(\zeta)$ gives the transformed equation

$$\zeta v'(\zeta) = (n-1)v(\zeta) + \left(\sum_{k=1}^\infty \pi_k \zeta^k\right) v(\zeta).$$

Repeating the transformation, we remove the term violating the condition. A modification of this approach works for systems of equations.

In our case the situation is different. Let us consider again the scalar case $m = 1$. Here the condition (3.42) is equivalent to the condition $\pi_0 \ne -x$ (the general solution of the equation $\pi_0 - \pi_0^q = [k]$ has the form $\pi_0 = -x + \xi$ where $\xi - \xi^{q^k} = 0$, that is either $\xi = 0$, or $|\xi| = 1$; the latter contradicts our assumption $|\pi_0| < 1$). If, on the contrary, $\pi_0 = -x$, then, as we saw in Theorem 3.8, $g(t) = 0$ for $|t| \le q^{-1}$, and the construction (3.43) does not make sense. On the other hand, a formal analog of the shearing transformation for this case is the substitution $u = \tau(v)$. However it is easy to see that v satisfies an equation with the same principal part, as the equation for u.

3.3.4. Euler-type equations. Classically, the Euler equation has the form
$$\zeta^m u^{(m)}(\zeta) + \beta_{m-1}\zeta^{m-1}u^{(m-1)}(\zeta) + \cdots + \beta_0 u(\zeta) = 0$$
where $\beta_0, \beta_1, \ldots \beta_{m-1} \in \mathbb{C}$. It can be reduced to a first-order linear system with a constant matrix. The solutions are linear combinations of functions of the form $\zeta^\lambda(\log \zeta)^k$. Of course, such functions have no direct $\mathbb{F}_q$-linear counterparts, and our study of the Euler-type equations will again be based on expansions in the Carlitz polynomials.

Let us consider a linear equation
$$\tau^m d^m u + b_{m-1}\tau^{m-1}d^{m-1}u + \cdots + b_0 u = 0 \tag{3.51}$$
where $b_0, b_1, \ldots, b_{m-1} \in \overline{K}_c$. In order to reduce (3.51) to a first-order system, it is convenient to set
$$\varphi_k = \tau^{k-1}d^{k-1}u, \quad k = 1, \ldots, m.$$
It is easy to prove by induction that $d\tau^{k-1} - \tau^{k-1}d = [k-1]^{1/q}\tau^{k-2}$. Therefore
$$\tau d\varphi_k = \tau^k d^k u + [k-1]\tau^{k-1}d^{k-1}u = \varphi_{k+1} + [k-1]\varphi_k,$$
$k = 1, \ldots, m-1$. Next, by (3.51),
$$\tau d\varphi_m = \tau^m d^m u + [m-1]\tau^{m-1}d^{m-1}u$$
$$= ([m-1] - b_{m-1})\varphi_m - b_{m-2}\varphi_{m-1} - \cdots - b_0\varphi_1.$$

Thus equation (3.51) can be written as a system
$$\tau d\varphi = B\varphi, \quad \varphi = (\varphi_1, \ldots, \varphi_m), \tag{3.52}$$

where

$$B = \begin{pmatrix} 0 & 1 & 0 & 0 & \cdots & 0 \\ 0 & [1] & 1 & 0 & \cdots & 0 \\ 0 & 0 & [2] & 1 & \cdots & 0 \\ 0 & 0 & 0 & [3] & \cdots & 0 \\ \cdots & \cdots & \cdots & \cdots & \cdots & \cdots \\ 0 & 0 & 0 & 0 & \cdots & 1 \\ -b_0 & -b_1 & -b_2 & -b_3 & \cdots & [m-1] - b_{m-1} \end{pmatrix}.$$

This time we cannot directly use the above results, since $|B| \geq 1$. However in some cases it is possible to proceed in a slightly different way.

Suppose that all the eigenvalues of the matrix B lie in the open disk $\{|\lambda| < 1\}$. Transforming B to its Jordan normal form we find that

$$B = X^{-1}(B_0 + N)X$$

where X is an invertible matrix, B_0 is a diagonal matrix, $|B_0| = \mu < 1$, N is nilpotent, that is $N^\varkappa = 0$ for some natural number $\varkappa$, and N commutes with B_0. If Ψ is a matrix solution of the system

$$\tau d\Psi = (B_0 + N)\Psi,$$

then $\Phi = X^{-1}\Psi X$ is a matrix solution of (3.52).

On the other hand, we can obtain Ψ just as in the case $N = 0$ considered in Section 3.3.2, writing

$$\Psi(t) = \sum_{i=0}^{\infty} c_i f_i(t), \tag{3.53}$$

$$c_i = \left\{ \prod_{j=0}^{i-1} (B_0 + N - [j]I_m) \right\} c_0. \tag{3.54}$$

Indeed, the product in (3.54) is the sum of the expressions $(-[j])^{\nu_1} B_0^{\nu_2} N^{\nu_3}$ where $\nu_1 + \nu_2 + \nu_3 = i$, $\nu_3 < \varkappa$. Therefore in (3.53)

$$|c_i| \leq |c_0| \cdot |N|^{\varkappa - 1} \left\{ \max(\mu, q^{-1}) \right\}^{i - \varkappa} \longrightarrow 0, \quad i \to \infty.$$

Let us consider in detail the case where $m = 2$. Our equation is

$$\tau^2 d^2 u + b_1 \tau du + b_0 u = 0. \tag{3.55}$$

Now we have the system (3.52) with

$$B = \begin{pmatrix} 0 & 1 \\ -b_0 & [1] - b_1 \end{pmatrix}.$$

The characteristic polynomial of B is $D_2(\lambda) = \lambda^2 + \lambda(b_1 - [1]) + b_0$, with the discriminant $\delta = (b_1 - [1])^2 - 4b_0$. We assume that the eigenvalues are such that $|\lambda_1|, |\lambda_2| < 1$. This condition is satisfied, for example, if $p \neq 2$, $|b_0| < 1$, $|b_1| < 1$.

The greatest common divisor of the first order minors of $B - \lambda I_2$ is 1. This means that B is diagonalizable if and only if $\lambda_1 \neq \lambda_2$, that is if $\delta \neq 0$ (see, e.g. [41]). In this case

$$B = X^{-1} \begin{pmatrix} \lambda_1 & 0 \\ 0 & \lambda_2 \end{pmatrix} X \qquad (3.56)$$

for some invertible matrix X, and our system has a matrix solution Φ, such that

$$X\Phi(t)X^{-1} = tI_2 + \sum_{n=1}^{\infty} \operatorname{diag}\left\{\prod_{j=0}^{n-1}(\lambda_1 - [j]), \prod_{j=0}^{n-1}(\lambda_2 - [j])\right\} f_n(t)$$

$$\stackrel{\text{def}}{=} \operatorname{diag}\{\psi_1(t), \psi_2(t)\}.$$

It is easy to see that $\psi_1(t)$ and $\psi_2(t)$ are solutions of equation (3.55). Indeed, if $X^{-1} = \begin{pmatrix} \xi_{11} & \xi_{12} \\ \xi_{21} & \xi_{22} \end{pmatrix}$, then

$$\Phi(t)X^{-1}\begin{pmatrix} 1 \\ 0 \end{pmatrix} = (\xi_{11}\psi_1(t), \xi_{21}\psi_2(t)),$$

and (for ψ_1) it is sufficient to show that $\xi_{11} \neq 0$. However $X^{-1}X = I_2$, and if $\xi_{11} = 0$, then writing $X = \begin{pmatrix} \chi_{11} & \chi_{12} \\ \chi_{21} & \chi_{22} \end{pmatrix}$ we find that $\xi_{12}\chi_{22} = 0$, $\xi_{12}\chi_{21} = 1$. At the same time, by (3.56), $\xi_{12}\lambda_2\chi_{21} = 0$, and $\xi_{12}\lambda_2\chi_{22} = 1$, and we come to a contradiction. Similar reasoning works for $\psi_2(t)$. It follows from the uniqueness of the Fourier–Carlitz expansion that ψ_1 and ψ_2 are linearly independent.

If $\lambda_1 = \lambda_2 \stackrel{\text{def}}{=} \lambda$, then B is similar to the Jordan cell

$$N = \begin{pmatrix} \lambda & 1 \\ 0 & \lambda \end{pmatrix}.$$

It is proved by induction that

$$\prod_{j=0}^{n-1}(N-[j]I_2) = \begin{pmatrix} \prod_{j=0}^{n-1}(\lambda-[j]) & \sum_{j=0}^{n-1}\prod_{\substack{0\le i\le n-1\\ i\ne j}}(\lambda-[i]) \\ 0 & \prod_{j=0}^{n-1}(\lambda-[j]) \end{pmatrix}.$$

In this case we have the following two linearly independent solutions of (3.55):

$$\psi_1(t) = t + \sum_{n=1}^{\infty}\left\{\prod_{j=0}^{n-1}(\lambda-[j])\right\}f_n(t),$$

$$\psi_2(t) = t + \sum_{n=1}^{\infty}\left\{\sum_{j=0}^{n-1}\prod_{\substack{0\le i\le n-1\\ i\ne j}}(\lambda-[i])\right\}f_n(t).$$

Thus, for the case of the eigenvalues from the disk $\{|\lambda|<1\}$, we have given an explicit construction of solutions for the Euler type equations.

3.4 Evolution equations

3.4.1. Classes of functions and operators. After the above basic properties of the Carlitz differential equations are established, a natural next step is to try to consider partial differential equations. However, here we encounter serious difficulties, some phenomena absent in the classical situation. For example, if we consider the natural action of the Carlitz derivatives d_s and d_t on an $\mathbb{F}_q$-linear monomial $f(s,t) = s^{q^m}t^{q^n}$, we notice immediately that $d_s^m f$ is not a polynomial, nor even a holomorphic function in t, if $m > n$ (since the action of d is not linear and involves taking the q-th root). Moreover, it follows from the relation $d\left(s^{q^m}\right) = [m]^{1/q}s^{q^{m-1}}$ and the commutation property $d\lambda = \lambda^{1/q}d$, where a scalar $\lambda \in \overline{K}_c$ is identified with the operator of multiplication by λ, that d_s and d_t do not commute even on monomials f with $m < n$.

It appears that nevertheless there is a class of partial differential operators (acting on an appropriate class of functions of several variables), which possess reasonable properties. Operators from this class contain the derivative d in only one distinguished variable, and the linear operator Δ in every other variable. Such operators d and Δ (in different variables) do not commute either but satisfy a simple commutation relation. Algebraic structures related to this class of operators will be considered in detail in Chapter 5.

Denote by $\mathcal{F}_{n+1}$ ($n \geq 1$) the set of all germs of functions of the form

$$u = \sum_{i_1=0}^{\infty} \cdots \sum_{i_n=0}^{\infty} \sum_{m=0}^{\min(i_1,\ldots,i_n)} c_{m,i_1,\ldots,i_n} s_1^{q^{i_1}} \cdots s_n^{q^{i_n}} \frac{z^{q^m}}{D_m} \qquad (3.57)$$

where $c_{m,i_1,\ldots,i_n} \in \overline{K}_c$ are such that all the series are convergent on some neighborhoods of the origin. Let $\hat{\mathcal{F}}_{n+1}$ be the set of polynomials from $\mathcal{F}_{n+1}$.

Below d will denote the Carlitz derivative in the variable z, while Δ_j will mean the difference operator Δ in the variable s_j. In the action of each operator d, Δ_j on a function from $\mathcal{F}_{n+1}$ (acting in a single variable) other variables are treated as scalars. Obviously, linear operators Δ_j commute with multiplications by scalars: $\Delta_j \lambda = \lambda \Delta_j$, $\lambda \in \overline{K}_c$, while $d\lambda = \lambda^{1/q} d$.

3.4.2. The Cauchy problem. Let us consider equations of the form

$$\{P(\Delta_1, \ldots, \Delta_n) + Q(\Delta_1, \ldots, \Delta_n) d\} u = 0 \qquad (3.58)$$

where P, Q are nonzero polynomials with coefficients from $\overline{K}_c$. We look for a solution $u \in \mathcal{F}_{n+1}$ of the form (3.57) satisfying an "initial condition"

$$\lim_{z \to 0} z^{-1} u(z, s_1, \ldots, s_n) = u_0(s_1, \ldots, s_n) \qquad (3.59)$$

where $u_0(s_1, \ldots, s_n)$ is an $\mathbb{F}_q$-linear holomorphic function on a neighborhood of the origin. The condition (3.59) (similar to the initial conditions for "ordinary" differential equations with Carlitz derivatives) means actually that the coefficients $c_{0,i_1,\ldots,i_n}$ of the solution (3.57) are prescribed for any $i_1, \ldots, i_n$.

Below we use the notation $[\infty] = -x$. Then $[n] \to [\infty]$, as $n \to \infty$.

Theorem 3.10 *Suppose that*

$$Q([i_1], \ldots, [i_n]) \neq 0 \quad \text{for all } i_1, \ldots, i_n = 0, 1, \ldots, \infty. \qquad (3.60)$$

Then the Cauchy problem (3.58)–(3.59) has a unique solution $u \in \mathcal{F}_{n+1}$.

Proof. It is easy to see that

$$\Delta_j \left(s^{q^{i_j}} \right) = \begin{cases} [i_j] s^{q^{i_j}}, & \text{if } i_j \neq 0; \\ 0, & \text{if } i_j = 0. \end{cases}$$

Differential equations 133

The identity $D_m = [m]D_{m-1}^q$ implies the relation

$$d\left(\frac{z^{q^m}}{D_m}\right) = \begin{cases} \frac{z^{q^{m-1}}}{D_{m-1}}, & \text{if } m \neq 0; \\ 0, & \text{if } m = 0. \end{cases}$$

Therefore for the function (3.57) we get

$$du = \sum_{i_1=1}^{\infty} \cdots \sum_{i_n=1}^{\infty} \sum_{m=1}^{\min(i_1,\ldots,i_n)} c_{m,i_1,\ldots,i_n}^{1/q} s_1^{q^{i_1}-1} \cdots s_n^{q^{i_n}-1} \frac{z^{q^{m-1}}}{D_{m-1}}$$

$$= \sum_{j_1=0}^{\infty} \cdots \sum_{j_n=0}^{\infty} \sum_{\nu=0}^{\min(j_1,\ldots,j_n)} c_{\nu+1,j_1+1,\ldots,j_n+1}^{1/q} s_1^{q^{j_1}} \cdots s_n^{q^{j_n}} \frac{z^{q^\nu}}{D_\nu}.$$

Next,

$$P(\Delta_1,\ldots,\Delta_n)u$$

$$= \sum_{i_1=0}^{\infty} \cdots \sum_{i_n=0}^{\infty} \sum_{m=0}^{\min(i_1,\ldots,i_n)} c_{m,i_1,\ldots,i_n} P([i_1],\ldots,[i_n]) s_1^{q^{i_1}} \cdots s_n^{q^{i_n}} \frac{z^{q^m}}{D_m}.$$

Writing a similar formula for $Q(\Delta_1,\ldots,\Delta_n)u$ and substituting all this into (3.58) we find that

$$\sum_{i_1=0}^{\infty} \cdots \sum_{i_n=0}^{\infty} \sum_{m=0}^{\min(i_1,\ldots,i_n)} \left\{ c_{m,i_1,\ldots,i_n} P([i_1],\ldots,[i_n]) \right.$$

$$\left. + c_{m+1,i_1+1,\ldots,i_n+1}^{1/q} Q([i_1],\ldots,[i_n]) \right\} s_1^{q^{i_1}} \cdots s_n^{q^{i_n}} \frac{z^{q^m}}{D_m} = 0$$

for arbitrary values of the variables. Hence, we come to the recursion

$$c_{m+1,i_1+1,\ldots,i_n+1} = -c_{m,i_1,\ldots,i_n}^q \left\{ \frac{P([i_1],\ldots,[i_n])}{Q([i_1],\ldots,[i_n])} \right\}^q,$$

$$m \leq \min(i_1,\ldots,i_n). \quad (3.61)$$

Since all the elements $c_{0,i_1,\ldots,i_n}$ $(i_1,\ldots,i_n = 0,1,2,\ldots)$ are given, from (3.61) we find all the coefficients of (3.57).

The set $\{[i], i = 0,1,\ldots,\infty\}$ is compact in K. Therefore the condition (3.60) implies the inequality

$$|Q([i_1],\ldots,[i_n])| \geq \mu > 0 \quad (3.62)$$

for all $i_1, \ldots, i_n = 0, 1, 2, \ldots, \infty$. Note also that $|[i]| = q^{-1}$ for any i, and

$$|c_{0,i_1,\ldots,i_n}| \leq Cr^{q^{i_1}+\cdots+q^{i_n}} \tag{3.63}$$

for some positive constants C and r, since the series for the initial condition

$$u_0(s_1,\ldots,s_n) = \sum_{i_1=0}^{\infty} \cdots \sum_{i_n=0}^{\infty} c_{0,i_1,\ldots,i_n} s_1^{q^{i_1}} \cdots s_n^{q^{i_n}}$$

converges near the origin.

By (3.62) and (3.63),

$$|c_{1,i_1+1,\ldots,i_n+1}| \leq C_1^q r^{q^{i_1+1}+\cdots+q^{i_n+1}}$$

(where $C_1 > 0$ does not depend on $i_1, \ldots, i_n$), thus

$$|c_{2,i_1+2,\ldots,i_n+2}| \leq C_1^{q^2+q} r^{q^{i_1+2}+\cdots+q^{i_n+2}},$$

and, by induction

$$|c_{l,i_1+l,\ldots,i_n+l}| \leq C_1^{q^l+q^{l-1}+\cdots+q} r^{q^{i_1+l}+\cdots+q^{i_n+l}}$$

for any $l \geq 0$. This means that for any $j_1, \ldots, j_n \geq l$,

$$|c_{l,j_1,\ldots,j_n}| \leq C_2^{q^l} r^{q^{j_1}+\cdots+q^{j_n}}, \quad C_2 > 0,$$

which, together with the equality

$$|D_m| = q^{-\frac{q^m-1}{q-1}},$$

implies the convergence of the series in (3.57) near the origin. ∎

Remark 3.2 It is easy to generalize Theorem 3.10 to the case of systems of equations, where P and Q are matrices whose elements are polynomials of $\Delta_1, \ldots, \Delta_n$. In this case, instead of (3.60) we have to require the invertibility of $Q([i_1], \ldots, [i_n])$ for all $i_1, \ldots, i_n = 0, 1, \ldots, \infty$. In an obvious way, this generalization covers also the case of a scalar equation of a higher order in d.

Equations (3.58) can be seen as function field analogs of classical evolution equations of mathematical physics. Specific examples of equation (3.58) related to hypergeometric functions will be considered in Chapter 4. In Chapter 5, we will consider modules of holonomic type corresponding to the general equations (3.58).

3.5 Comments

The general theory of the Carlitz differential equations (Section 3.1) was initiated by the author [63]. An important motivation was to cover, by general theorems, the analogs of the classical hypergeometric equations introduced by Thakur [109, 110]. See Chapter 4 for an exposition of this subject.

The material given in the subsequent sections is taken from [66] (strongly nonlinear equations), [64] (regular singularity), and [71] (evolution equations).

4
Special functions

This chapter is devoted to some $\mathbb{F}_q$-linear special functions defined or interpreted in terms of the Carlitz differential equations. In particular, we consider hypergeometric functions, analogs of the Bessel functions, polylogarithms and K-binomial coefficients. We discuss overconvergence properties of some special functions.

4.1 Hypergeometric functions

4.1.1. Hypergeometric equations. Let us consider an evolution equation (in the sense of Section 3.4.2)
$$\{P(\Delta_1,\ldots,\Delta_n) + Q(\Delta_1,\ldots,\Delta_n)d\}\, u = 0 \tag{4.1}$$
with $n \geq \max(r,s)$, $r, s \in \mathbb{N}$,
$$P(t_1,\ldots,t_n) = \prod_{i=1}^{r}(t_i - a_i),$$
$$Q(t_1,\ldots,t_n) = \prod_{j=1}^{s}(t_j - b_j),$$
where $a_i, b_j \in \overline{K}_c$, and the elements b_j do not coincide with any of the elements $[\nu]$, $\nu = 0, 1, \ldots, \infty$. The condition (3.60) is satisfied, and the Cauchy problem for equation (4.1) is well-posed.

Let us specify the initial condition (in terms of prescribing the values of $c_{0,i_1,\ldots,i_n}$ in the expansion (3.57) of a solution) as follows:
$$c_{0,0,\ldots,0} = 1, \quad c_{0,i_1,\ldots,i_n} = 0$$

for all other values of $i_1, \ldots, i_n$. Then $c_{m,i_1,\ldots,i_n} = 0$ for all sets of indices $(m, i_1, \ldots, i_n)$ except those with $m = i_1 = \ldots = i_n$. Denote $\sigma_m = c_{m,m,\ldots,m}$. By (3.61), we find that

$$\sigma_1 = \left\{ \frac{(-1)^r \prod_{i=1}^r a_i}{(-1)^s \prod_{j=1}^s b_j} \right\}^q,$$

$$\sigma_2 = \left\{ \frac{(-1)^r \prod_{i=1}^r a_i}{(-1)^s \prod_{j=1}^s b_j} \right\}^{q^2} \left\{ \frac{\prod_{i=1}^r ([1] - a_i)}{\prod_{j=1}^s ([1] - b_j)} \right\}^q,$$

and, by induction, after rearranging the factors we get

$$\sigma_m = \frac{\prod_{i=1}^r ([0] - a_i)^{q^m} ([1] - a_i)^{q^{m-1}} \cdots ([m-1] - a_i)^q}{\prod_{j=1}^s ([0] - b_j)^{q^m} ([1] - b_j)^{q^{m-1}} \cdots ([m-1] - b_j)^q}, \quad m = 1, 2, \ldots.$$
(4.2)

For $a \in \overline{K}_c$, denote $\langle a \rangle_0 = 1$,

$$\langle a \rangle_m = ([0] - a)^{q^m} ([1] - a)^{q^{m-1}} \cdots ([m-1] - a)^q, \quad m \geq 1 \qquad (4.3)$$

(of course, $[0] = 0$, but we maintain the symbol $[0]$ to have an orderly notation). The Pochhammer-type symbol $\langle \cdot \rangle_m$ satisfies the recurrence

$$\langle a \rangle_{m+1} = ([m] - a)^q \langle a \rangle_m^q. \qquad (4.4)$$

It follows from (4.2) that the solution $u(z; t_1, \ldots, t_n)$ of the above Cauchy problem is given by the formula

$$u(z; t_1, \ldots, t_n) = {}_rF_s(a_1, \ldots, a_r; b_1, \ldots, b_s; t_1 \cdots t_n z) \qquad (4.5)$$

where ${}_rF_s$ is the *hypergeometric function*

$${}_rF_s(a_1, \ldots, a_r; b_1, \ldots, b_s; z) = \sum_{m=0}^{\infty} \frac{\langle a_1 \rangle_m \cdots \langle a_r \rangle_m}{\langle b_1 \rangle_m \cdots \langle b_s \rangle_m} \frac{z^{q^m}}{D_m}. \qquad (4.6)$$

The series (4.6) has a positive radius of convergence since $b_1, \ldots, b_s$ do not coincide with any of the elements $[\nu]$, $\nu = 0, 1, \ldots, \infty$ (such parameters will be called *admissible*).

Let h_m be the coefficients of the power series (4.5), that is

$$h_m = \frac{\langle a_1 \rangle_m \cdots \langle a_r \rangle_m}{\langle b_1 \rangle_m \cdots \langle b_s \rangle_m D_m}, \quad m = 0, 1, 2, \ldots.$$

Since $\dfrac{D_{m+1}}{D_m^q} = [m+1] = \left(x^{q^m} - x^{q^{-1}}\right)^q = ([m] - [-1])^q$, we find that

$$\frac{h_{m+1}}{h_m^q} = \left\{ \frac{([m] - a_1) \cdots ([m] - a_r)}{([m] - b_1) \cdots ([m] - b_s)([m] - [-1])} \right\}^q. \quad (4.7)$$

The identity (4.7) means that the ratio $\dfrac{h_{m+1}}{h_m^q}$ is the q-th power of a rational function of $[m]$, which is a clear analog of the basic property of the classical hypergeometric function. Note that any rational function of $[m]$ may appear in (4.7), except those for which (4.7) does not make sense.

It follows from (4.4) that $\Delta_i u = \Delta u$ where $\Delta = \tau d$ is the Carlitz difference operator in the variable z. Therefore $_rF_s$ satisfies the equation

$$\left\{ \prod_{i=1}^r (\Delta - a_i) - \left(\prod_{j=1}^s (\Delta - b_j) \right) d \right\} {}_rF_s = 0. \quad (4.8)$$

In particular, for the Gauss-like hypergeometric function $_2F_1$ we have

$$\{(\Delta - a)(\Delta - b) - (\Delta - c)d\}{}_2F_1 = 0.$$

Substituting $\Delta = \tau d$ and using the commutation relations

$$d\tau - \tau d = [1]^{1/q} I, \quad \Delta \tau - \tau \Delta = [1]\tau,$$

we can rewrite this equation in the form

$$\left\{ \tau(1-\tau)d^2 - \left(c - ([1]^{1/q} + a + b)\tau \right) d - ab \right\} {}_2F_1 = 0, \quad (4.9)$$

resembling the classical hypergeometric equation.

If $b \neq [\nu]$, $\nu = 0, 1, \ldots, \infty$, then, in particular, $|b + x| \geq \mu > 0$, whence

$$|b - [\nu]| = \left| (b+x) - x^{q^\nu} \right| = |b + x| \geq \mu$$

for large values of ν. This means that

$$|b - [\nu]| \geq \mu_1 > 0, \quad \nu = 0, 1, 2, \ldots,$$

so that

$$|\langle b \rangle_\nu| \geq \mu_1^{q^\nu + q^{\nu-1} + \cdots + q} = \mu_1^{\frac{q^{\nu+1}-1}{q-1} - 1} = \mu_1^{-(q-1)^{-1} - 1} \left(\mu_1^{\frac{q}{q-1}} \right)^{q^\nu}.$$

Therefore, if $|z|$ is small enough, then the series (4.6) converges uniformly with respect to the parameters $a_i \in \overline{K}_c$ and $b_j \in \overline{K}_c \setminus \{[\nu], \nu = 0, 1, \ldots, \infty\}$ on any compact set. Thus, the function $_rF_s$ is a locally analytic function of its parameters.

All the above constructions make sense if, instead of the field K, we consider a completion of the global field $\mathbb{F}_q(x)$ with respect to any of its places. Only minor changes are needed in the case of a finite place corresponding to an irreducible polynomial $\pi \in \mathbb{F}_q[x]$ (the field K is obtained if $\pi(x) = x$). In the case of the "infinite" valuation some additional assumptions are needed to guarantee convergence of the series defining the hypergeometric series (4.6).

4.1.2. Thakur's hypergeometric functions. The first definitions of hypergeometric functions over K (or, globally, over completions of $\mathbb{F}_q(x)$) were proposed by Thakur [109, 110].

For $\alpha_i, \beta_i \in \mathbb{Z}$, such that the series below makes sense, Thakur's hypergeometric function is defined as

$$_rF_s(\alpha_1, \ldots, \alpha_r; \beta_1, \ldots, \beta_s; z) = \sum_{n=0}^{\infty} \frac{(\alpha_1)_n \cdots (\alpha_r)_n}{(\beta_1)_n \cdots (\beta_s)_n D_n} z^{q^n} \qquad (4.10)$$

where

$$(\alpha)_n = \begin{cases} D_{n+\alpha-1}^{q^{-(\alpha-1)}}, & \text{if } \alpha \geq 1; \\ (-1)^{n-\alpha} L_{-\alpha-n}^{-q^n}, & \text{if } \alpha \leq 0, n \leq -\alpha; \\ 0, & \text{if } \alpha \leq 0, n > -\alpha. \end{cases} \qquad (4.11)$$

If $a = [-\alpha]$, $\alpha \in \mathbb{Z}$, then

$$([m] - a)^q = \left(x^{q^m} - x^{q^{-\alpha}}\right)^q = \left(x^{q^{m+\alpha}} - x\right)^{q^{-(\alpha-1)}} = [m+\alpha]^{q^{-(\alpha-1)}},$$

and it is easy to check that the Pochhammer–Thakur symbols (4.11) satisfy the same recurrence (4.4) as the above symbols (4.3). The normalization $\langle a \rangle_0 = 1$ is different from (4.11) and resembles the classical one. Therefore, if

$$a_i = [-\alpha_i], b_j = [-\beta_j] \ (i = 1, \ldots, r; j = 1, \ldots, s), \quad \alpha_i \in \mathbb{Z}, \beta_j \in \mathbb{N},$$

then the functions (4.10) and (4.6) coincide up to a change of variable $z \Rightarrow \rho z$ where ρ depends on all the parameters but does not depend on z. The appropriate counterparts of equations (4.8) and (4.9) were among the first nontrivial examples of Carlitz differential equations.

Special functions 141

As for the rationality property (4.7), Thakur's hypergeometric function corresponds to the case of rational functions with zeros and poles of the form $[\nu]$, $\nu \in \mathbb{Z}$.

Note that, in addition to the function (4.10), Thakur considered other possible solutions of the hypergeometric equation. In some cases such solutions can be nonholomorphic, even strongly singular [64].

Apart from (4.10), in [109, 110] Thakur introduced the second analog of the hypergeometric function, with parameters a_i, b_j from completions of $\mathbb{F}_q(x)$, and the expressions $e_n(a)$ instead of the symbols $(a)_n$. Both kinds of hypergeometric function satisfy various identities resembling the classical ones; however no connections of Thakur's second analog with the Carlitz differential equations have been found so far.

4.1.3. Contiguous relations. Among many identities for classical hypergeometric functions, a special role belongs to relations between contiguous functions, that is the hypergeometric functions $_2F_1$ whose parameters differ by ± 1 [9]. For Thakur's hypergeometric functions, analogs of the contiguous relations were found in [109, 110, 111]. Here we present analogs of the contiguous relations for our more general situation. Of 15 possible relations, we give, just as Thakur did, only two, which is sufficient to demonstrate specific features of the function field case; other relations can be obtained in a similar way. Note that the specializations of our identities for the case of parameters like $[-\alpha]$, $\alpha \in \mathbb{Z}$, are slightly different from those in [109, 110, 111], due to a different normalization of our Pochhammer-type symbols.

Our first task is to find an appropriate counterpart of the shift by 1 for parameters from $\overline{K}_c$. Denote

$$T_1(a) = (a - [1])^{1/q}, \quad a \in \overline{K}_c. \tag{4.12}$$

If $a = [-\alpha]$, $\alpha \in \mathbb{Z}$, then

$$T_1([-\alpha]) = ([-\alpha] - [1])^{1/q} = x^{q^{-\alpha-1}} - x = [-\alpha - 1],$$

so that the transformation T_1 indeed extends the unit shift of integers. The inverse transformation is given by

$$T_{-1}(a) = a^q + [1], \quad a \in \overline{K}_c. \tag{4.13}$$

Theorem 4.1 *The following identities for the Pochhammer-type symbol* and the Gauss-type hypergeometric function are valid for every $m \in \mathbb{N}$ and any admissible parameters from $\overline{K}_c$:

$$\langle T_1(a) \rangle_m = a^{-q^m}(a - [m])\langle a \rangle_m, \quad a \neq 0; \tag{4.14}$$

$$\langle a \rangle_{m+1} = -a^{q^{m+1}} \langle T_1(a) \rangle_m^q; \tag{4.15}$$

$$\langle T_{-1}(a) \rangle_m = -([1] + a^q)^{q^m} \langle a \rangle_{m-1}^q; \tag{4.16}$$

$$\langle T_{-1}(a) \rangle_m = -\frac{([1] + a^q)^{q^m}}{([m-1] - a)^q} \langle a \rangle_m; \tag{4.17}$$

$$_2F_1(T_1(a), b; c; az) - {}_2F_1(a, T_1(b); c; bz) = (a - b){}_2F_1(a, b; c; z); \tag{4.18}$$

$$_2F_1(a, b; c; z) - {}_2F_1(a, b; c; z)^q + (c^q - b^q){}_2F_1(a, b; T_1(c); c^{-1}z)^q$$
$$- (a^q + [1]){}_2F_1(T_{-1}(a), b; c; (a^q + [1])^{-1}z) = 0. \tag{4.19}$$

Proof. Substituting (4.12) into (4.3) and using the fact that $[\nu] + [1]^{1/q} = x^{q^\nu} - x^{1/q} = [\nu + 1]^{1/q}$, we get

$$\langle T_1(a) \rangle_m = ([1] - a)^{q^{m-1}}([2] - a)^{q^{m-2}} \cdots ([m] - a), \tag{4.20}$$

which implies (4.14). If we raise both sides of (4.20) to the power q and compare the resulting identity with (4.4), we come to (4.15). The proofs of (4.16) and (4.17) are similar, based on the identity $[\nu] - [1] = [\nu - 1]^q$.

Using (4.14) we find that

$$_2F_1(T_1(a), b; c; az) = \sum_{m=0}^{\infty}(a - [m])\frac{\langle a \rangle_m \langle b \rangle_m}{\langle c \rangle_m D_m}z^{q^m},$$

$$_2F_1(a, T_1(b); c; bz) = \sum_{m=0}^{\infty}(b - [m])\frac{\langle a \rangle_m \langle b \rangle_m}{\langle c \rangle_m D_m}z^{q^m},$$

which implies (4.18). Similarly, if we write down all the terms involved in (4.19) and use the identities (4.4), (4.14), (4.15), and (4.17), after rather lengthy but quite elementary calculations we verify the required identity (4.19). ∎

4.2 Analogs of the Bessel functions and Jacobi polynomials

4.2.1. The Bessel-type functions. Let

$$J_n(t) = \sum_{k=0}^{\infty}(-1)^k \frac{t^{q^{n+k}}}{D_{n+k}D_k^{q^n}}, \quad n = 0, 1, 2, \ldots \tag{4.21}$$

The series (4.21) converges for $|t| < q^{-\frac{2}{q-1}}$.

Computing the action of the Carlitz difference operator Δ on J_n we find that
$$\Delta^2 J_n(t) - [n]\Delta J_n(t) + \tau J_n(t) = 0. \tag{4.22}$$

The equation (4.22) can be reduced to a first order system (3.41) with regular singularity.

Writing also
$$J_{-n}(t) = (-1)^n \{J_n(t)\}^{q^{-n}}$$

we get, for all $n \in \mathbb{Z}$, the identity
$$\Delta^{(r)} J_n(t) = \tau^r J_{n-r}(t), \quad r = 0, 1, 2, \ldots,$$

and the recurrence relation
$$J_{n+1}(t) - [n]J_n(t) + \tau J_{n-1}(t) = 0.$$

See [24] for some identities involving the second linearly independent solution of the equation (4.22), as well as an analog of the modified Bessel function. The latter is defined by the series
$$I_n(t) = \sum_{k=0}^{\infty} \frac{t^{q^{n+k}}}{D_{n+k} D_k^{q^n}}, \quad n = 0, 1, 2, \ldots.$$

The functions I_n and J_n are connected as follows. Let $\gamma \in \overline{K}$ be a solution of the equation $\gamma^{q-1} = -1$. Then $\gamma^{q^r} = (-1)^r \gamma$ for any natural number r, so that $I_n(\gamma t) = (-1)^n \gamma J_n(t)$. On the other hand, I_n can be expressed via Thakur's hypergeometric function; see [109].

Just as the Carlitz exponential e_C, the function J_n becomes an entire function over K_∞. See [24].

Let us consider some expansions related to the Bessel–Carlitz functions J_n.

Proposition 4.2 *If $|t| < q^{-\frac{2}{q-1}}$, then*
$$\frac{t^{q^n}}{D_n} = \sum_{l=0}^{\infty} \frac{1}{L_l^{q^n}} J_{l+n}(t). \tag{4.23}$$

Proof. We have

$$\left|\frac{1}{D_{n+k}D_k^{q^n}}\right| = q^{\frac{2q^{n+k}-q^n-1}{q-1}} \leq q^{\frac{2q^{n+k}}{q-1}},$$

so that for $|t| < q^{-\frac{2}{q-1}-\varepsilon}$, $\varepsilon > 0$,

$$|J_m(t)| \leq \sup_{k \geq 0} q^{-\varepsilon q^{m+k}} = q^{-\varepsilon q^m}. \tag{4.24}$$

Since $\left|\dfrac{1}{L_l^{q^n}}\right| = q^{lq^n}$, we find from the inequality (4.24) with $m = l + n$ that the series (4.23) converges on the above region of t.

The right-hand side of (4.23) equals

$$\sum_{l=0}^\infty \frac{1}{L_l^{q^n}} \sum_{\nu=0}^\infty (-1)^\nu \frac{t^{q^{l+n+\nu}}}{D_{l+n+\nu}D_\nu^{q^{l+n}}} = \sum_{k=0}^\infty \frac{t^{q^{k+n}}}{D_{k+n}} \left\{\sum_{r=0}^k \frac{(-1)^{k-r}}{L_r D_{k-r}^{q^r}}\right\}^{q^n} = \frac{t^{q^n}}{D_n},$$

as desired, due to the identity (1.63). ∎

We will obtain further expansions after we introduce some polynomials having an independent interest.

4.2.2. A class of polynomials. The polynomials $P_m^{(n,k)}(t)$, $m, n, k \in \mathbb{Z}_+$, are defined as follows:

$$P_m^{(n,k)}(t) = \sum_{r=0}^m (-1)^{m-r} \begin{bmatrix} m \\ r \end{bmatrix}^{q^{n+k}} \frac{D_{r+n+k}}{D_{r+k}^{q^n}} t^{q^{r+n+k}}, \tag{4.25}$$

where $\begin{bmatrix} m \\ r \end{bmatrix} = \dfrac{D_m}{D_r D_{m-r}^{q^r}}$ appears also in the explicit formula (1.14) for the additive Carlitz polynomials e_i. Comparing (4.25) and (1.14) we see that for $n = 0$,

$$P_m^{(0,k)}(t) = (e_m(t))^{q^k}.$$

Moreover,

$$P_m^{(n,k)}(t) = \Delta^{(n)} (e_m(t))^{q^{n+k}}.$$

There are some other interesting special cases of the polynomials $P_m^{(n,k)}$. For example, if we allow the value $k = -n$, $n \leq m$, we find that

$$P_m^{(n,-n)}(t) = \frac{D_m}{D_{m-n}^{q^n}} e_{m-n}^{q^n}(t).$$

Some nice identities are obtained for the modified polynomials
$$\widetilde{P}_m^{(n,k)}(t) = \left\{ P_m^{(n,k)}(t) \right\}^{q^{-n}}.$$

It can be verified directly that
$$\Delta \widetilde{P}_m^{(n,k)}(t) = \left\{ \widetilde{P}_m^{(n+1,k-1)}(t) \right\}^q.$$

The above connection with the Carlitz polynomials e_m leads to various identities involving them together with $P_m^{(n,k)}$ or $\widetilde{P}_m^{(n,k)}$ (see [24]).

Let us prove an expansion formula involving $J_k(t)$ and $P_m^{(n,k)}(t)$.

Proposition 4.3 *If* $s, t \in K$, $|s| \leq 1$, $|t| < q^{-\frac{2}{q-1}}$, *then*

$$J_k^{q^n}(st) = \sum_{m=0}^{\infty} \frac{(-1)^m}{D_m^{q^{n+k}}} J_{m+k+n}(t) P_m^{(n,k)}(s). \qquad (4.26)$$

Proof. By the definition of J_k,
$$J_k^{q^n}(st) = \sum_{r=0}^{\infty} (-1)^r \frac{(st)^{q^{r+k+n}}}{D_r^{q^{k+n}} D_{r+k}^{q^n}}.$$

Using (4.23) we get
$$J_k^{q^n}(st) = \sum_{r=0}^{\infty} (-1)^r \frac{D_{r+k+n}}{D_r^{q^{k+n}} D_{r+k}^{q^n}} s^{q^{r+k+n}} \sum_{l=0}^{\infty} \frac{1}{L_l^{q^{r+k+n}}} J_{r+l+k+n}(t)$$
$$= \sum_{m=0}^{\infty} J_{m+k+n}(t) \sum_{r=0}^{\infty} (-1)^r \frac{D_{r+k+n}}{D_r^{q^{k+n}} D_{r+k}^{q^n} L_{m-r}^{q^{r+k+n}}} s^{q^{r+k+n}}.$$

Now the identity (4.26) follows from the definition of $P_m^{(n,k)}(s)$. ∎

See [24] for further identities containing the above functions and polynomials.

4.3 Polylogarithms

4.3.1. A logarithm-like function. We begin with an analog of the function $-\log(1-t)$ defined via the equation

$$(1-\tau)du(t) = t, \quad t \in K, \qquad (4.27)$$

a counterpart of the classical equation $(1 - t)u'(t) = 1$ (note that the function $i(t) = t$ is a unit element in the composition ring).

Let us look for an $\mathbb{F}_q$-linear holomorphic solution

$$u(t) = \sum_{n=0}^{\infty} a_n t^{q^n} \qquad (4.28)$$

of equation (4.27). We have

$$du(t) = \sum_{n=1}^{\infty} a_n^{1/q}[n]^{1/q} t^{q^{n-1}}.$$

Substituting into (4.27) we find that

$$\sum_{j=0}^{\infty} \left(a_{j+1}^{1/q}[j+1]^{1/q} - a_j[j] \right) t^{q^j} = t,$$

and equation (4.27) is satisfied if and only if a_0 is arbitrary, $a_1 = [1]^{-1}$,

$$a_{j+1} = a_j^q \frac{[j]^q}{[j+1]}, \quad j \geq 1,$$

By induction, we find that $a_j = [j]^{-1}$.

Let $l_1(t)$ be the solution (4.28) of equation (4.27) with $a_0 = 0$. Then

$$l_1(t) = \sum_{n=1}^{\infty} \frac{t^{q^n}}{[n]}. \qquad (4.29)$$

The series in (4.29) converges if $t \in K$, $|t| \leq q^{-1}$.

The function $l_1(t)$ is clearly different from the Carlitz logarithm $\log_C$ (see Chapter 1). From the composition ring viewpoint, $\log_C(t)$ is an analog of e^{-t}, though in other respects it is a valuable analog of the logarithm. By the way, another possible analog of the logarithm is a continuous (on O) solution of the equation $\Delta u(t) = t$, an analog of $tu'(t) = 1$, such that $u(1) = 0$. By the identity (1.36), $u = \mathcal{D}_1$, the first hyperdifferentiation; see Chapter 1.

Now we consider continuous nonholomorphic extensions of l_1. We will use the following simple lemma.

Lemma 4.4 *Consider the equation*

$$z^q - z = \xi, \quad \xi \in \overline{K}_c. \qquad (4.30)$$

If $|\xi| = 1$, then for all the solutions $z_1, \ldots, z_q$ of equation (4.30), $|z_1| = \ldots = |z_q| = 1$. If $|\xi| < 1$, then there exists a unique solution z_1 of equation (4.30) with $|z_1| = |\xi|$. This solution can be written as

$$z_1 = -\sum_{j=0}^{\infty} \xi^{q^j}. \tag{4.31}$$

For all other solutions we have $|z_j| = 1$, $j = 2, \ldots, q$.

Proof. Let $|\xi| = 1$. If some solution z_j of equation (4.30) is such that $|z_j| < 1$, then the ultra-metric inequality would imply $|\xi| < 1$. If $|z_j| > 1$, then $|z_j|^q > |z_j|$, so that $|\xi| = |z_j|^q > 1$, and we again come to a contradiction.

Now suppose that $|\xi| < 1$. Then the series in (4.31) converges and defines a solution of (4.30), such that $|z_1| = |\xi|$. All other solutions are obtained by adding elements of $\mathbb{F}_q$ to z_1. Therefore $|z_2| = \ldots = |z_q| = 1$. ■

Theorem 4.5 *Equation (4.27) has exactly q continuous solutions on O coinciding with (4.29) as $|t| \leq q^{-1}$. These solutions have the expansions in the normalized Carlitz polynomials*

$$u = \sum_{i=0}^{\infty} c_i f_i \tag{4.32}$$

where the coefficient c_1 is an arbitrary solution of the equation

$$c_1^q - c_1 + 1 = 0, \tag{4.33}$$

higher coefficients are found from the relation

$$c_n = \sum_{j=0}^{\infty} (c_{n-1}[n-1])^{q^{j+1}}, \quad n \geq 2, \tag{4.34}$$

and the coefficient c_0 is determined by the relation

$$c_0 = \sum_{i=1}^{\infty} (-1)^{i+1} \frac{c_i}{L_i}, \tag{4.35}$$

Proof. Writing equation (4.27) in the form $du(t) - \Delta u(t) = t$ and using the relations (1.21) and (3.35) we find that

$$\sum_{i=0}^{\infty} \left(c_{i+1}^{1/q} - c_{i+1} - c_i[i] \right) f_i(t) = f_0(t), \quad t \in O.$$

This is equivalent to equation (4.33) for c_1 and the relations

$$c_n^q - c_n + [n-1]^q c_{n-1}^q = 0, \quad n \geq 2. \tag{4.36}$$

By Lemma 4.4, there are q solutions of equation (4.33), and for any of them $|c_1| = 1$. For $n = 2$, we find from Lemma 4.4 and the equality $|[1]| = q^{-1}$ that the corresponding equation (4.36) has the solution (4.34) with $|c_2| = q^{-q}$ and $q-1$ other solutions with absolute value 1. Choosing at each subsequent step the solution (4.34) we obtain the sequence c_n, such that

$$|c_n| \leq q^{-q^{n-1}}, \quad n \geq 2, \tag{4.37}$$

so that $|c_n| \to 0$, and the series (4.32) indeed determines a continuous $\mathbb{F}_q$-linear function on O. Since

$$f_i(t) = \sum_{j=0}^{i} (-1)^{i-j} \frac{1}{D_j L_{i-j}^{q^j}} t^{q^j},$$

we see that

$$\lim_{t \to 0} \frac{f_i(t)}{t} = \frac{(-1)^i}{L_i}.$$

Therefore, if we choose c_0 according to (4.35), then our solution u is such that

$$\lim_{t \to 0} t^{-1} u(t) = 0. \tag{4.38}$$

Note that $|L_n| = q^{-n}$, so that the series in (4.35) is convergent.

By Corollary 2.22, it follows from (4.37) that u is locally analytic; specifically, it is analytic on any ball of radius q^{-1}. Thus it can be represented for $|t| \leq q^{-1}$ by convergent power series (4.28), in which $a_0 = 0$ by (4.38). Therefore $u(t) = l_1(t)$ for $|t| \leq q^{-1}$, as desired.

Any other continuous solution of equation (4.27) on O is obtained inevitably by the same procedure, but with $|c_1| = \ldots = |c_N| = 1$, $|c_n| < 1$, if $n \geq N+1$, with some $N > 1$, and with some $c_0 \in \overline{K}_c$. In this case by Lemma 4.4, $|c_{N+1}| = q^{-q}$ and, by induction,

$$|c_{N+l}| = q^{-\left(q^l + q^{(l-1)} + \cdots + q\right)}. \tag{4.39}$$

Indeed, this was shown above for $l = 1$. If (4.39) is true for some l, then

$$|c_{N+l+1}| = |c_{N+l}|^q q^{-q} = q^{-\left(q^{l+1} + q^l + \cdots + q^2 + q\right)},$$

Special functions 149

and we have proved (4.39) for any $l \geq 1$. It will be convenient to write (4.39) in the form
$$|c_{N+l}| = q^{-\alpha_l}, \quad \alpha_l = q^l + q^{(l-1)} + \cdots + q. \tag{4.40}$$
Suppose that our solution coincides with the series (4.29) for $|t| \leq q^{-1}$. By $\mathbb{F}_q$-linearity, this means the analyticity of the solution on any ball of radius q^{-1}. Using a specialization of Corollary 2.21 for the $\mathbb{F}_q$-linear case we find that
$$-\log_q |c_n| - \sum_{i=2}^{\infty} q^{n-i} \longrightarrow \infty \quad \text{as } n \to \infty,$$
that is
$$-\log_q |c_n| - \frac{q^{n-1}}{q-1} \longrightarrow \infty \quad \text{as } n \to \infty.$$
In particular, we obtain that
$$\alpha_l - \frac{q^{N+l-1}}{q-1} \longrightarrow \infty \quad \text{as } l \to \infty. \tag{4.41}$$
However by (4.40),
$$\alpha_l = \frac{q^{l+1} - q}{q-1},$$
which contradicts (4.41), since $N \geq 2$. ∎

In fact, continuous solutions which satisfy (4.35) and have coefficients c_n of the form (4.34), but starting from some larger values of n, are also extensions of the function (4.29), but from smaller balls.

Below we denote by $l_1(t)$ a fixed solution of equation (4.27) on O coinciding with (4.29) for $|t| \leq q^{-1}$, as described in Theorem 4.5.

4.3.2. Polylogarithms. The polylogarithms $l_n(t)$ are defined recursively by the equations
$$\Delta l_n = l_{n-1}, \quad n \geq 2, \tag{4.42}$$
which agree with the classical ones $t l_n'(t) = l_{n-1}(t)$. If we look for analytic $\mathbb{F}_q$-linear solutions of (4.42), such that $t^{-1} l_n(t) \to 0$ as $t \to 0$, we obtain easily by induction that
$$l_n(t) = \sum_{j=1}^{\infty} \frac{t^{q^j}}{[j]^n}, \quad |t| \leq q^{-1}. \tag{4.43}$$

In order to find continuous extensions of l_n onto O, we consider the Carlitz expansions

$$l_n = \sum_{i=0}^{\infty} c_i^{(n)} f_i, \quad n = 2, 3, \ldots \tag{4.44}$$

Consider first the dilogarithm l_2. We have

$$\Delta l_2 = \sum_{i=0}^{\infty} \left(c_{i+1}^{(2)} + [i] c_i^{(2)} \right) f_i,$$

so that

$$c_{i+1}^{(2)} + [i] c_i^{(2)} = c_i, \quad i = 0, 1, 2, \ldots, \tag{4.45}$$

where c_i are the coefficients described in Theorem 4.5. The recursion (4.45) leaves $c_0^{(2)}$ arbitrary and determines all other coefficients in a unique way:

$$c_n^{(2)} = (-1)^n L_{n-1} \sum_{j=n}^{\infty} (-1)^j \frac{c_j}{L_j}, \quad n \geq 1. \tag{4.46}$$

Indeed, the series in (4.46) is convergent, since c_n satisfies the estimate (4.37), while $|L_n| = q^{-n}$. For $n = 1$ the equality (4.46) means, due to (4.35), that $c_1^{(2)} = c_0$, which coincides with (4.45) for $i = 0$. If (4.46) is proved for some n, then

$$c_{n+1}^{(2)} = c_n - [n] c_n^{(2)} = c_n + (-1)^{n+1} L_n \sum_{j=n}^{\infty} (-1)^j \frac{c_j}{L_j}$$

$$= (-1)^{n+1} L_n \sum_{j=n+1}^{\infty} (-1)^j \frac{c_j}{L_j},$$

as desired.

We have

$$\left| \frac{c_j}{L_j} \right| = q^{j - q^{j-1}},$$

so that for $n > 1$,

$$\left| \sum_{j=n}^{\infty} (-1)^j \frac{c_j}{L_j} \right| \leq \sup_{j \geq n} \left| \frac{c_j}{L_j} \right| \leq q \sup_{j \geq n} \left(q^{j-1} \right) q^{-q^{j-1}}.$$

The function $z \mapsto zq^{-z}$ is monotone decreasing for $z \geq 1$. Therefore

$$\left|\sum_{j=n}^{\infty}(-1)^j \frac{c_j}{L_j}\right| \leq q^n \cdot q^{-q^{n-1}}, \quad n > 1,$$

so that by (4.46)

$$\left|c_n^{(2)}\right| \leq q \cdot q^{-q^{n-1}}, \quad n > 1.$$

Using Corollary 2.22 as above, we find that l_2 is analytic on all balls of the radius q^{-1}. If we choose $c_0^{(2)}$ in such a way that

$$c_0^{(2)} = \sum_{i=1}^{\infty}(-1)^{i+1}\frac{c_i^{(2)}}{L_i},$$

the solution (4.44) of equation (4.42) with $n = 2$ is a continuous extension of the dilogarithm l_2 given by the series (4.43) with $n = 2$.

Repeating the above reasoning for each n, we come to the following result.

Theorem 4.6 *For each $n \geq 2$, there exists a unique continuous $\mathbb{F}_q$-linear solution of equation (4.42) coinciding for $|t| \leq q^{-1}$ with the polylogarithm (4.43). The solution is given by the Carlitz expansion (4.44) with*

$$\left|c_i^{(n)}\right| \leq C_n q^{-q^{i-1}}, \quad i > 1, \ C_n > 0,$$

$$c_0^{(n)} = \sum_{i=1}^{\infty}(-1)^{i+1}\frac{c_i^{(n)}}{L_i}.$$

The above construction and properties of polylogarithms can be implemented also in the global field setting, for any finite place of $\mathbb{F}_q(x)$. See [67] for the details.

4.3.3. Zeta function. We define $\zeta(t)$, $t \in K$, setting $\zeta(0) = 0$,

$$\zeta(x^{-n}) = l_n(1), \quad n = 1, 2, \ldots,$$

and then using the fractional derivative of Section 2.1.5 and setting

$$\zeta(t) = \left(\Delta^{(\theta_0 + \theta_1 x + \cdots)} l_n\right)(1), \quad n = 1, 2, \ldots,$$

if $t = x^{-n}(\theta_0 + \theta_1 x + \cdots)$, $\theta_j \in \mathbb{F}_q$. The correctness of this definition follows from Proposition 2.9. It is clear that ζ is a continuous $\mathbb{F}_q$-linear function on K with values in $\overline{K}_c$.

In particular, we have
$$\zeta(x^m) = \left(\Delta^{m+1}l_1\right)(1), \quad m = 0, 1, 2, \ldots.$$

The above definition is of course inspired by the classical polylogarithm relation
$$\left(z\frac{d}{dz}\right)\sum_{n=1}^{\infty}\frac{z^n}{n^s} = \sum_{n=1}^{\infty}\frac{z^n}{n^{s-1}}.$$

Let us write down some relations for "special values" $\zeta(x^n)$, $n \in \mathbb{N}$. Let us consider the expansion of $l_n(t)$ in the sequence of hyperdifferentiations. By Theorem 1.12,
$$l_n(t) = \sum_{i=0}^{\infty}\left(\Delta^i l_n\right)(1)\mathcal{D}_i(t).$$

Therefore
$$l_n(t) = \sum_{i=0}^{\infty}\zeta(x^{-n+i})\mathcal{D}_i(t), \quad n \in \mathbb{N},\ t \in O. \tag{4.47}$$

In particular, combining (4.47) and (4.43) we get
$$\sum_{j=1}^{\infty}\frac{t^{q^j}}{[j]^n} = \sum_{i=0}^{\infty}\zeta(x^{-n+i})\mathcal{D}_i(t), \quad |t| \leq q^{-1}.$$

Let us consider the double sequence $A_{n,r} \in K$, $A_{n,1} = (-1)^{n-1}L_{n-1}$,
$$A_{n,r} = (-1)^{n+r}L_{n-1}\sum_{0<i_1<\ldots<i_{r-1}<n}\frac{1}{[i_1][i_2]\cdots[i_{r-1}]}, \quad r \geq 2.$$

This sequence appears in the expansion of a hyperdifferentiation $\mathcal{D}_r$ in the normalized Carlitz polynomials (see (1.31)), as well as in the expression [54] of the Carlitz difference operators $\Delta^{(n)}$ via the iterations Δ^n:
$$\Delta^{(n)} = \sum_{r=1}^{n}A_{n,r}\Delta^r, \quad n \geq 1. \tag{4.48}$$

For coefficients of the expansion (4.44) we have $c_i^{(n)} = \left(\Delta^{(i)}l_n\right)(1)$, $i \geq 1$ (see Theorem 1.8), and by (4.48),
$$c_i^{(n)} = \sum_{r=1}^{i}A_{i,r}\left(\Delta^r l_n\right)(1) = \sum_{r=1}^{i}A_{i,r}\zeta(x^{r-n}). \tag{4.49}$$

Since $c_0^{(n)} = \zeta(x^{-n})$, we have (see Theorems 4.5 and 4.6)

$$\zeta(x^{-n}) = \sum_{i=1}^{\infty}(-1)^{i+1}L_i^{-1}\sum_{r=1}^{i}A_{i,r}\zeta(x^{r-n}). \quad (4.50)$$

The identity (4.50) may be seen as a distant relative of Riemann's functional equation for the classical zeta function.

Since $\mathcal{D}_r(t)$ is not differentiable (Section 1.3.1), the interpretation of the sequence $\{A_{i,r}\}$ given in (1.31) shows, by Lemma 2.2, that $L_i^{-1}A_{i,r} \twoheadrightarrow 0$ as $i \to \infty$. Thus it is impossible to change the order of summation in (4.50).

Finally, consider the coefficients of the expansion (4.32) for l_1. As in (4.49), we have an expression

$$c_i = \sum_{r=1}^{i} A_{i,r}\zeta(x^{r-1}).$$

By Theorem 4.5, for $i \geq 2$ we have

$$c_i = \sum_{j=0}^{\infty}(z_i)^{q^j}, \quad z_i = c_{i-1}^q[i-1]^q \in \overline{K}_c. \quad (4.51)$$

The series in (4.51) may be seen as an analog of $\sum_{j} j^{-z}$. This analogy becomes clearer if, for a fixed $z \in \overline{K}_c$, $|z| < 1$, we consider the set S of all convergent power series $\sum_{n=1}^{\infty} z^{q^{j_n}}$ corresponding to sequences $\{j_n\}$ of natural numbers. Let us introduce the multiplication $\otimes$ in S setting $z^{q^i} \otimes z^{q^j} = z^{q^{ij}}$ and extending the operation distributively (for a similar construction in the framework of q-analysis in characteristic 0 see [80]). Denoting by $\prod_{p}^{\otimes}$ the product in S of elements indexed by prime numbers we obtain in a standard way the identity

$$c_i = \prod_{p}^{\otimes} \sum_{n=0}^{\infty}(z_i)^{q^{p^n}}$$

(the infinite product is understood as a limit of the partial products in the topology of Ω_x), an analog of the Euler product formula.

The above polylogarithms and zeta function are different from the existing objects playing an essential part at the function field arithmetic (see [45, 94, 111]) and having no relations to the Carlitz differential equations. At present the author is unaware of any arithmetic applications, though hopes for them to emerge. This hope is based on the fact that the above ζ is purely an object of the characteristic p arithmetic, while Goss's zeta function is defined initially on $\mathbb{Z}$ and interpolated onto $\mathbb{Z}_p$.

4.4 K-binomial coefficients

4.4.1. Absolute values.
Let us consider the K-binomial coefficients

$$\binom{k}{m}_K = \frac{D_k}{D_m D_{k-m}^{q^m}}, \quad 0 \leq m \leq k, \tag{4.52}$$

introduced in the positive characteristic version of umbral calculus (Section 2.2). A straightforward calculation shows that

$$\left|\binom{k}{m}_K\right| = 1, \quad 0 \leq m \leq k.$$

Since $\binom{k}{m}_K \in \mathbb{F}_q(x)$, it is natural to consider also other places of $\mathbb{F}_q(x)$ corresponding to monic irreducible polynomials $\pi \in \mathbb{F}_q[x]$ (see Section 1.5). Let $\delta = \deg \pi$, and denote by $|\cdot|_\pi$ the absolute value on $\mathbb{F}_q(x)$ corresponding to π.

Below we prove that the expression (4.52) belongs to the ring of integers not only for the field K, but for any finite place of the global function field $\mathbb{F}_q(x)$. Together with Section 2.2, this property supports the case for considering the expressions (4.52) as "proper" analogs of the classical binomial coefficients. For other analogs see [111].

Proposition 4.7 *For any monic irreducible polynomial $\pi \in \mathbb{F}_q[x]$, the K-binomial coefficients (4.52) satisfy the inequality*

$$\left|\binom{k}{m}_K\right|_\pi \leq 1, \quad 0 \leq m \leq k.$$

Proof. First we compute $|D_m|_\pi$. It follows from Lemma 2.13 of [76] that

$$|[i]|_\pi = \begin{cases} q^{-\delta}, & \text{if } \delta \text{ divides } i, \\ 1, & \text{otherwise.} \end{cases}$$

Writing $m = j\delta + i$, with $i, j \in \mathbb{Z}_+$, $0 \le i < \delta$, we find that

$$|D_m|_\pi = |[j\delta]|_\pi^{q^i} |[(j-1)\delta]|_\pi^{q^{\delta+i}} \ldots |[\delta]|_\pi^{q^{(j-1)\delta+i}}$$

$$= \left\{ q^{-\delta} \cdot \left(q^{-\delta}\right)^{q^\delta} \cdot \ldots \cdot \left(q^{-\delta}\right)^{q^{(j-1)\delta}} \right\}^{q^i}$$

$$= \left\{ \left(q^{-\delta}\right)^{1+q^\delta + \cdots + q^{(j-1)\delta}} \right\}^{q^i} = q^{-\delta q^i \frac{q^{j\delta}-1}{q^\delta - 1}}.$$

(we already made similar computations in the proof of Lemma 1.19).

Similarly we can write $k - m = \varkappa\delta + \lambda$, with $\varkappa, \lambda \in \mathbb{Z}_+$, $0 \le \lambda < \delta$, and get that

$$|D_{k-m}| = q^{-\delta q^\lambda \frac{q^{\varkappa\delta}-1}{q^\delta-1}}.$$

If $i + \lambda < \delta$, then we obtain a similar representation for k simply by adding those for m and $k - m$, so that

$$\log_q \left| \binom{k}{m} \right|_{K|\pi}$$

$$= -\frac{\delta}{q^\delta - 1} \left\{ q^{i+\lambda} \left(q^{(j+\varkappa)\delta} - 1 \right) - q^i \left(q^{j\delta} - 1 \right) - q^\lambda \left(q^{\varkappa\delta} - 1 \right) q^{j\delta+i} \right\}$$

$$= -\frac{\delta}{q^\delta - 1} q^i \left(1 + q^{\lambda+j\delta} - q^\lambda - q^{j\delta} \right)$$

$$= -\frac{\delta}{q^\delta - 1} q^i \left(q^\lambda - 1 \right) \left(q^{j\delta} - 1 \right) \le 0.$$

If $i + \lambda \ge \delta$, then $k = (j + \varkappa + 1)\delta + \nu$ where $0 \le \nu = i + \lambda - \delta < \delta$. In this case

$$\log_q \left| \binom{k}{m} \right|_{K|\pi}$$

$$= -\frac{\delta}{q^\delta - 1} \left\{ q^\nu \left(q^{(j+\varkappa+1)\delta} - 1 \right) - q^i \left(q^{j\delta} - 1 \right) - q^\lambda \left(q^{\varkappa\delta} - 1 \right) q^{j\delta+i} \right\}$$

$$= -\frac{\delta}{q^\delta - 1} \left(q^i + q^{\lambda+j\delta+i} - q^{i+j\delta} - q^\nu \right) < 0,$$

since $\nu < i + \lambda$. ∎

Below we will use only the valuation with $\pi(x) = x$, that is, as above, consider the field K.

4.4.2. Identities. Let us derive, for the K-binomial coefficients (4.52), analogs of the classical Pascal and Vandermonde identities.

Proposition 4.8 *The identity*
$$\binom{k}{m}_K = \binom{k-1}{m-1}_K^q + \binom{k-1}{m}_K^q D_m^{q-1} \qquad (4.53)$$
holds, if $0 \leq m \leq k$ and it is assumed that $\binom{k}{-1}_K = \binom{k-1}{k}_K = 0.$

Proof. Let $e_m(t) = D_m f_m(t)$ be the "nonnormalized" Carlitz polynomials. They satisfy the main K-binomial identity (see (2.30))
$$e_k(st) = \sum_{m=0}^{k} \binom{k}{m}_K e_m(s) \{e_{k-m}(t)\}^{q^m}, \qquad (4.54)$$
which holds, for example, for any $s, t \in \mathbb{F}_q[x]$.

It is known (see Proposition 1.7) that
$$e_k = e_{k-1}^q - D_{k-1}^{q-1} e_{k-1}. \qquad (4.55)$$

Let us rewrite the left-hand side of (4.54) in accordance with (4.55), and apply to each term the identity (4.54) with $k-1$ substituted for k. We have
$$e_{k-1}^q(st) = \sum_{i=0}^{k-1} \binom{k-1}{i}_K^q e_i^q(s) e_{k-i-1}^{q^{i+1}}(t).$$

By (4.55), $e_i^q = e_{i+1} + D_i^{q-1} e_i$, $e_{k-i-1}^q = e_{k-i} + D_{k-i-1}^{q-1} e_{k-i-1}$, whence
$$e_{k-1}^q(st) = \sum_{j=1}^{k} \binom{k-1}{j-1}_K^q e_j(s) e_{k-j}^{q^j}(t) + \sum_{i=0}^{k-1} \binom{k-1}{i}_K^q D_i^{q-1} e_i(s) e_{k-i}^{q^i}(t)$$
$$+ \sum_{i=0}^{k-1} \binom{k-1}{i}_K^q D_i^{q-1} D_{k-i-1}^{q^i(q-1)} e_i(s) e_{k-i-1}^{q^i}(t).$$

Note that
$$\binom{k-1}{i}_K^q D_i^{q-1} D_{k-i-1}^{q^i(q-1)} = D_{k-1}^{q-1} \binom{k-1}{i}_K. \qquad (4.56)$$

Indeed, the left-hand side of (4.56) equals

$$\frac{D_{k-1}^q}{D_i^q D_{k-i-1}^{q^{i+1}}} D_i^{q-1} D_{k-i-1}^{q^{i+1}-q^i} = \frac{D_{k-1}}{D_i D_{k-i-1}^{q^i}} D_{k-1}^{q-1}$$

and coincides with the right-hand side. Therefore the last sum in the expression for $e_{k-1}^q(st)$ equals

$$D_{k-1}^{q-1} \sum_{i=0}^{k-1} \binom{k-1}{i}_K^{q^i} e_i(s) e_{k-i-1}^{q^i}(t) = D_{k-1}^{q-1} e_{k-1}(st).$$

Using (4.55) again we find that

$$e_k(st) = \sum_{i=0}^{k} \binom{k-1}{i-1}_K^q e_i(s) e_{k-i}^{q^i}(t) + \sum_{i=0}^{k} \binom{k-1}{i}_K^q D_i^{q-1} e_i(s) e_{k-i}^{q^i}(t),$$

and the comparison with (4.54) yields

$$\sum_{m=0}^{k} \left\{ \binom{k}{m}_K - \binom{k-1}{m-1}_K^q - \binom{k-1}{m}_K^q D_m^{q-1} \right\} e_m(s) e_{k-m}^{q^m}(t) = 0$$

for any s, t.

Since the Carlitz polynomials are linearly independent, we obtain that

$$\left\{ \binom{k}{m}_K - \binom{k-1}{m-1}_K^q - \binom{k-1}{m}_K^q D_m^{q-1} \right\} e_{k-m}^{q^m}(t) = 0$$

for any t, and it remains to note that $e_{k-m}(t) \neq 0$ if $t \in \mathbb{F}_q[x]$, $\deg t \geq k$, by the definition of the Carlitz polynomials. ∎

More generally, we have the following Vandermonde-type identity. Let k, m be integers, $0 \leq m \leq k$.

Proposition 4.9 *Define $c_{li}^{(m)} \in K$ by the recurrence relation*

$$c_{l+1,i}^{(m)} = c_{l,i-1}^{(m)} + c_{li}^{(m)} D_{m-i}^{q-1} \tag{4.57}$$

and the initial conditions $c_{li}^{(m)} = 0$ for $i < 0$ and $i > l$, $c_{00}^{(m)} = 1$. Then, for any $l \leq m$,

$$\binom{k}{m}_K = \sum_{i=0}^{l} c_{li}^{(m)} \binom{k-l}{m-i}_K^{q^l}. \tag{4.58}$$

Proof. The identity (4.58) is trivial for $l = 0$. Suppose it has been proved for some l. Let us transform the right-hand side of (4.58) using the identity (4.53). Then we have

$$\binom{k}{m}_K = \sum_{i=0}^{l} c_{li}^{(m)} \binom{k-l-1}{m-i-1}_K^{q^{l+1}} + \sum_{i=0}^{l} c_{li}^{(m)} \binom{k-l-1}{m-i}_K^{q^{l+1}} D_{m-i}^{q-1}$$

$$= \sum_{j=1}^{l+1} c_{l,j-1}^{(m)} \binom{k-l-1}{m-j}_K^{q^{l+1}} + \sum_{i=0}^{l} c_{li}^{(m)} \binom{k-l-1}{m-i}_K^{q^{l+1}} D_{m-i}^{q-1}.$$

Since we assume that $c_{l,-1}^{(m)} = c_{l,l+1}^{(m)} = 0$, the summation in both the above sums can be performed from 0 to $l+1$. Using (4.57) we obtain the required identity (4.58) with $l+1$ substituted for l. ∎

4.4.3. A Carlitz differential equation. Now we consider a function $f \in \mathcal{F}_2$ (see (3.57)) associated with the K-binomial coefficients, that is

$$f(s,t) = \sum_{k=0}^{\infty} \sum_{m=0}^{k} \binom{k}{m}_K s^{q^m} t^{q^k}. \tag{4.59}$$

Proposition 4.10 *The function (4.59) satisfies the equation*

$$d_s f(s,t) = \Delta_t f(s,t) + [1]^{1/q} f(s,t). \tag{4.60}$$

Proof. Let us compute $d_s f$. We have

$$d_s f(s,t) = \sum_{k=1}^{\infty} \sum_{m=1}^{k} \binom{k}{m}_K^{1/q} [m]^{1/q} s^{q^{m-1}} t^{q^{k-1}}$$

$$= \sum_{\nu=0}^{\infty} \sum_{\mu=0}^{\nu} \binom{\nu+1}{\mu+1}_K^{1/q} [\mu+1]^{1/q} s^{q^\mu} t^{q^\nu}.$$

Using Proposition 4.8 we find that $d_s f = \Sigma_1 + \Sigma_2$ where

$$\Sigma_1 = \sum_{\nu=0}^{\infty} \sum_{\mu=0}^{\nu} \binom{\nu}{\mu}_K [\mu+1]^{1/q} s^{q^\mu} t^{q^\nu},$$

$$\Sigma_2 = \sum_{\nu=0}^{\infty} \sum_{\mu=0}^{\nu} \binom{\nu}{\mu+1}_K [\mu+1]^{1/q} D_{\mu+1}^{1-q^{-1}} s^{q^\mu} t^{q^\nu}.$$

Note that
$$[\mu+1]^{1/q} = \left(x^{q^{\mu+1}} - x\right)^{1/q} = \left(x^{q^\mu} - x\right) + (x^q - x)^{1/q} = [\mu] + [1]^{1/q},$$
so that
$$\Sigma_1 = \sum_{\nu=0}^{\infty} \sum_{\mu=0}^{\nu} \binom{\nu}{\mu}_K [\mu] s^{q^\mu} t^{q^\nu} + [1]^{1/q} f(s,t). \tag{4.61}$$

Next, we have
$$\binom{\nu}{\mu+1}_K [\mu+1]^{1/q} D_{\mu+1}^{1-q^{-1}} = \frac{D_\nu}{D_{\mu+1} D_{\nu-\mu-1}^{q^{\mu+1}}} D_{\mu+1} \left(\frac{[\mu+1]}{D_{\mu+1}}\right)^{1/q}$$
$$= \frac{D_\nu}{D_\mu D_{\nu-\mu-1}^{q^{\mu+1}}},$$

and also
$$D_{\nu-\mu-1}^q = \frac{1}{[\nu-\mu]}[\nu-\mu] D_{\nu-\mu-1}^q = \frac{D_{\nu-\mu}}{[\nu-\mu]},$$

whence
$$D_{\nu-\mu-1}^{q^{\mu+1}} = \frac{D_{\nu-\mu}^{q^\mu}}{[\nu-\mu]^{q^\mu}}.$$

Therefore
$$\Sigma_2 = \sum_{\nu=0}^{\infty} \sum_{\mu=0}^{\nu} \binom{\nu}{\mu}_K [\nu-\mu]^{q^\mu} s^{q^\mu} t^{q^\nu}.$$

As above, $[\nu-\mu]^{q^\mu} = \left(x^{q^{\nu-\mu}} - x\right)^{q^\mu} = [\nu] - [\mu]$, so that
$$\Sigma_2 = \sum_{\nu=0}^{\infty} \sum_{\mu=0}^{\nu} \binom{\nu}{\mu}_K ([\nu] - [\mu]) s^{q^\mu} t^{q^\nu}.$$

Together with (4.61), this implies (4.60). ∎

4.5 Overconvergence properties

4.5.1. The background. The idea of overconvergence is among the basic ones in contemporary p-adic analysis. In contrast to analysis over $\mathbb{R}$ and $\mathbb{C}$, the power series for principal special functions over $\mathbb{Q}_p$ or $\mathbb{C}_p$ converge only on finite disks or annuli. For example, the exponential series

$\exp(t) = \sum_{n=0}^{\infty} \frac{t^n}{n!}$, $t \in \mathbb{C}_p$, converges if and only if $|t|_p < p^{-1/(p-1)}$ (see [35] or [90]).

At the same time, for many special functions there exist some expressions combining their values in various points (usually connected by the Frobenius power $t \mapsto t^p$), for which the corresponding power series converge on wider regions. The simplest example is the Dwork exponential

$$\theta(t) = \exp(\pi(t - t^p)) \tag{4.62}$$

where π is a root of the equation $z^{p-1} + p = 0$. The power series for $\theta(t)$, in the variable t, converges for $|t|_p < p^{\frac{p-1}{p^2}}$ (> 1), though the formula (4.62) is not valid outside the unit disk. The special value $\theta(1)$ is a primitive p-th root of unity.

Other examples involve the exponential function of q-analysis [5], some hypergeometric functions [34], polylogarithms [28], and many others. The overconvergent functions usually satisfy equations possessing special algebraic properties called the Frobenius structures (see [5, 89]).

In this section we consider the overconvergence phenomena in the positive characteristic situation, that is for $\mathbb{F}_q$-linear functions on subsets of the field K. In particular, using the Carlitz exponential e_C we construct an analog of the Dwork exponential and prove its overconvergence, consider the overconvergence problems for some other special functions. These problems are much simpler than those in the characteristic zero case. The reason is that those functions satisfy differential equations with the Carlitz derivatives; the difference structure of the latter leads immediately to overconvergence properties of some linear combinations of solutions.

4.5.2. The Dwork–Carlitz exponential. Let σ be an arbitrary solution of the equation $z^{q-1} = -x$. Let us consider the function

$$E(t) = e_C(\sigma(t - t^q)), \tag{4.63}$$

defined initially for $|t| < q^{-\frac{1}{q-1}}$ (recall that we denote by $|\cdot|$ also the extension of the absolute value from K onto $\overline{K}_c$). Note that, in spite of the formal resemblance, the formulas for the Dwork exponential (4.62) and "the Dwork–Carlitz exponential" (4.63) have a quite different meaning – the function $\theta(t)$ is a multiplicative combination of values of the classical exponential, while $E(t)$ is an additive combination of values of the Carlitz exponential. This difference from classical overconvergence theory appears also in some other examples given below.

Special functions

From the definition of e_C, after a simple transformation we find that

$$E(t) = \sigma t + \sum_{n=1}^{\infty} \left(\frac{\sigma^{q^n}}{D_n} - \frac{\sigma^{q^{n-1}}}{D_{n-1}} \right) t^{q^n}. \tag{4.64}$$

In order to investigate the convergence of the series (4.64), we have to study the structure of elements D_n.

Proposition 4.11 *For any* $n \geq 1$,

$$\left| D_n - (-1)^n x^{1+q+\cdots+q^{n-1}} \right| \leq q^{-\frac{q^n-1}{q-1} - (q-1)q^{n-1}}. \tag{4.65}$$

Proof. We will prove that

$$\left| D_n - (-1)^n x^{1+q+\cdots+q^{n-1}} \right| \leq q^{-l_n} \tag{4.66}$$

where the sequence $\{l_n\}$ is determined by the requrrence

$$l_n = q l_{n-1} + 1, \quad l_1 = q. \tag{4.67}$$

Indeed, if $n = 1$, then $D_1 = x^q - x$, so that $|D_1 + x| = q^{-q}$. Suppose that we have proved (4.66) for some value of n. We have

$$\left| D_n^q - (-1)^n x^{q+q^2 \cdots + q^n} \right| \leq q^{-q l_n},$$

whence

$$\left| [n+1] D_n^q - (-1)^n [n+1] x^{q+q^2 \cdots + q^n} \right| \leq q^{-(q l_n + 1)}.$$

Since $D_{n+1} = [n+1] D_n^q$, we find that

$$\left| \left(D_{n+1} - (-1)^{n+1} x^{1+q+\cdots+q^n} \right) - (-1)^n x^{q+q^2 \cdots + q^{n+1}} \right| \leq q^{-l_{n+1}}. \tag{4.68}$$

It is easy to check that

$$l_n = \frac{q^n - 1}{q - 1} + (q-1) q^{n-1} \tag{4.69}$$

satisfies (4.67); the expression (4.69) can also be deduced from a general formula for a solution of a difference equation; see [42].

On the other hand,

$$\left| (-1)^n x^{q+q^2 \cdots + q^{n+1}} \right| = q^{-\frac{q^{n+2}-q}{q-1}}, \tag{4.70}$$

and, by a simple computation,

$$\frac{q^{n+2}-q}{q-1} - l_{n+1} = q^n - 1 > 0, \quad n \geq 1. \tag{4.71}$$

It follows from (4.68), (4.70), (4.71), and the ultra-metric property of the absolute value, that

$$\left| D_{n+1} - (-1)^{n+1} x^{1+q+\cdots+q^n} \right| \leq q^{-l_{n+1}},$$

which proves the inequalities (4.66) and (4.65) for any n. ∎

Now we can prove the overconvergence of $E(t)$.

Proposition 4.12 *The series in (4.64) converges for $|t| < \rho$, where $\rho = q^{\frac{q-1}{q^2}} > 1$. In particular, $E(1) = \lim_{n \to \infty} \frac{\sigma^{q^n}}{D_n}$ is defined, and $E(1) = \sigma$.*

Proof. Let us write

$$\frac{\sigma^{q^n}}{D_n} - \frac{\sigma^{q^{n-1}}}{D_{n-1}} = \frac{\sigma^{q^{n-1}}}{D_{n-1}} \left(\frac{\sigma^{q^n - q^{n-1}} D_{n-1}}{D_n} - 1 \right).$$

We have $\sigma^{q^n - q^{n-1}} = -x^{q^{n-1}}$, so that

$$\frac{\sigma^{q^n}}{D_n} - \frac{\sigma^{q^{n-1}}}{D_{n-1}} = -\frac{\sigma^{q^{n-1}}}{D_{n-1} D_n} \left(x^{q^{n-1}} D_{n-1} + D_n \right)$$

$$= -\frac{\sigma^{q^{n-1}}}{D_{n-1} D_n} \left\{ x^{q^{n-1}} \left(D_{n-1} - (-1)^{n-1} x^{1+\cdots+q^{n-2}} \right) \right.$$

$$\left. + \left(D_n - (-1)^n x^{1+\cdots+q^{n-1}} \right) \right\}.$$

If $n \geq 2$, then by Proposition 4.11,

$$\left| x^{q^{n-1}} \left(D_{n-1} - (-1)^{n-1} x^{1+\cdots+q^{n-2}} \right) \right| \leq q^{-\left(q^{n-1} + \frac{q^{n-1}-1}{q-1} + (q-1)q^{n-2} \right)},$$

$$\left| D_n - (-1)^n x^{1+\cdots+q^{n-1}} \right| \leq q^{-\frac{q^n-1}{q-1} - (q-1)q^{n-1}}.$$

Comparing the right-hand sides we check that the first of them is bigger; therefore

$$\left| \frac{\sigma^{q^n}}{D_n} - \frac{\sigma^{q^{n-1}}}{D_{n-1}} \right| \leq q^{-\frac{q^{n-1}}{q-1}} \cdot q^{-\frac{q^{n-1}-1}{q-1}} \cdot q^{-\frac{q^n-1}{q-1}} \cdot q^{-\left(q^{n-1} + \frac{q^{n-1}-1}{q-1} + (q-1)q^{n-2} \right)},$$

so that
$$\left|\frac{\sigma^{q^n}}{D_n} - \frac{\sigma^{q^{n-1}}}{D_{n-1}}\right| \leq q^{-\frac{q^{n-2}(q-1)^2+1}{q-1}}, \quad n \geq 2. \tag{4.72}$$

For $n = 1$, we get
$$\left|\frac{\sigma^q}{D_1} - \sigma\right| = |\sigma|\left|\frac{-x}{[1]} - 1\right| = |\sigma|\left|\frac{x^q}{[1]}\right|,$$

whence
$$\left|\frac{\sigma^q}{D_1} - \sigma\right| \leq q^{-\frac{1}{q-1} - (q-1)}. \tag{4.73}$$

It follows from (4.72) that the series in (4.64) converges for $|t| < \rho$. For $t = 1$, we obtain that
$$E(1) = \sigma + \sum_{n=1}^{\infty}\left(\frac{\sigma^{q^n}}{D_n} - \frac{\sigma^{q^{n-1}}}{D_{n-1}}\right) = \lim_{n\to\infty}\frac{\sigma^{q^n}}{D_n}. \tag{4.74}$$

Note that
$$\left|\frac{\sigma^{q^n}}{D_n}\right| = q^{-\frac{1}{q-1}}$$

for all values of n. Now
$$E(1)^q = \lim_{n\to\infty}\frac{\sigma^{q^{n+1}}}{D_n^q} = \lim_{n\to\infty}[n+1]\frac{\sigma^{q^{n+1}}}{D_{n+1}} = \lim_{n\to\infty}[n]\frac{\sigma^{q^n}}{D_n} = -xE(1)$$

because
$$\left|\frac{x^{q^n}\sigma^{q^n}}{D_n}\right| = q^{-\frac{1}{q-1} - q^n} \longrightarrow 0,$$

as $n \to \infty$.

By (4.74), $E(1) \neq 0$, so that $E(1)^{q-1} = -x$, thus $E(1)$ satisfies the same equation as σ. All the solutions of this equation are obtained by multiplying σ by nonzero elements $\xi \in \mathbb{F}_q$. Therefore $E(1) = \sigma\xi$, $\xi \in \mathbb{F}_q$, $\xi \neq 0$. If $\xi \neq 1$, then
$$|E(1) - \sigma| = |(1-\xi)\sigma| = |\sigma| = q^{-\frac{1}{q-1}}. \tag{4.75}$$

On the other hand, by (4.74),
$$|E(1) - \sigma| \leq \sup_{n \geq 1}\left|\frac{\sigma^{q^n}}{D_n} - \frac{\sigma^{q^{n-1}}}{D_{n-1}}\right|,$$

and we see that (4.75) contradicts (4.72) and (4.73). ∎

It is interesting that the special value $\sigma = E(1)$, just as the special value of the Dwork exponential in the characteristic 0 case, generates a cyclotomic extension of the function field (related in this case to the Carlitz module); see [94].

4.5.3. Examples of overconvergence. The Carlitz exponential e_C satisfies the simplest equation $de_C = e_C$, so that

$$e_C(t)^q + xe_C(t) = e_C(xt). \tag{4.76}$$

The right-hand side of (4.76) obviously converges on a wider disk than e_C itself (note that, in contrast to the p-adic case, $E(t)$ is not known to satisfy a homogeneous equation with the Carlitz derivative).

Similarly, the Bessel–Carlitz function $J_n(t)$ satisfies the identity $\Delta J_n = J_{n-1}^q$, so that

$$J_{n-1}^q(t) + xJ_n(t) = J_n(xt), \tag{4.77}$$

and we have an overconvergence for the right-hand side of (4.77). In this sense equations with the Carlitz derivatives may be seen themselves as analogs of the Frobenius structures of p-adic analysis.

The next two examples (of an essentially similar nature) are just a little more complicated.

Consider the polylogarithms

$$l_n(t) = \sum_{j=1}^{\infty} \frac{t^{q^j}}{[j]^n}, \quad n = 1, 2, \ldots \tag{4.78}$$

The series in (4.78) converges for $|t| < 1$. In Section 4.3, we constructed their continuous extensions to the disk $\{t \in K : |t| \leq 1\}$. Here we give the following overconvergence result resembling Coleman's theorem [28] about classical polylogarithms.

Proposition 4.13 *The power series for the function $L_n(t) = l_n(t) - l_n(t^q)$ converges for $|t| < q^{1/q}$.*

Proof. By a simple transformation, we get

$$L_n(t) = \frac{t^q}{[1]^n} + \sum_{j=2}^{\infty} \left(\frac{1}{[j]^n} - \frac{1}{[j-1]^n} \right) t^{q^j}. \tag{4.79}$$

We have,
$$\frac{1}{[j]^n} - \frac{1}{[j-1]^n} = \frac{([j-1]-[j])\left([j]^{n-1}+[j-1][j]^{n-2}+\cdots+[j-1]^{n-1}\right)}{[j]^n[j-1]^n}.$$

For any $j \geq 2$, $|[j]| = q^{-1}$, $|[j-1]-[j]| = \left|x^{q^{j-1}} - x^{q^j}\right| = q^{-q^{j-1}}$, so that

$$\left|\frac{1}{[j]^n} - \frac{1}{[j-1]^n}\right| \leq q^{-q^{j-1}+n-1},$$

and the convergence radius of the series (4.79) equals $q^{1/q}$. ∎

Let us consider the hypergeometric function

$$F(a,b;c;t) = \sum_{n=0}^{\infty} \frac{\langle a \rangle_n \langle b \rangle_n}{\langle c \rangle_n D_n} t^{q^n}, \qquad (4.80)$$

where $a, b, c \in \overline{K}_c$, $c \notin \{[0], [1], \ldots, [\infty]\}$, $[\infty] = -x$. For the notation see Section 4.1.

If $|a| = |b| = |c| = 1$, then the disk of convergence of the series (4.80) is $\left\{ t \in \overline{K}_c : |t| < q^{-\frac{1}{q-1}} \right\}$. In this case $|T_1(a)| = |T_1(b)| = |T_1(c)| = 1$.

Proposition 4.14 *The identity*

$$\tau F\left(T_1(a), T_1(b); T_1(c); \frac{ab}{c}t\right) - xF(a,b;c;t) = -F(a,b;c;xt) \qquad (4.81)$$

holds for any values of the variable and parameters, such that all the terms of (4.81) make sense. In particular, if $|a| = |b| = |c| = 1$, then the right-hand side of (4.81) is overconvergent, that is the series for the right-hand side converges for $|t| < q^{1-\frac{1}{q-1}}$ (>1).

Proof. Changing the index of summation we find that

$$\tau F(T_1(a), T_1(b); T_1(c); z) = \sum_{n=0}^{\infty} \frac{\langle T_1(a) \rangle_{n-1}^q \langle T_1(b) \rangle_{n-1}^q}{\langle T_1(c) \rangle_{n-1}^q D_{n-1}^q} t^{q^n}$$

for any z from the convergence disk. Using the identity (4.15) and the fact that $D_n = [n]D_{n-1}^q$ we get

$$\tau F(T_1(a), T_1(b); T_1(c); z) = -\sum_{n=0}^{\infty} \frac{\langle a \rangle_n \langle b \rangle_n [n]}{\langle c \rangle_n D_n} \left(\frac{ab}{c}\right)^{-q^n} z^{q^n}$$

(note that $[0] = 0$), which implies (4.81). ∎

4.6 Comments

The hypergeometric functions (4.6) were introduced by the author [71], as an application of the equations of evolution type (Section 3.4.2). A decisive motivation was to extend Thakur's definition [109, 110] of hypergeometric functions in such a way that the parameters would belong to $\overline{K}_c$, just as the argument and the values.

The Bessel–Carlitz functions and the associated polynomials were defined and studied by Carlitz [24]. See also [103].

The definitions and results regarding polylogarithms and the zeta function are taken from [67]. In this book we do not touch on the much richer theory of Goss's zeta function; see [45, 111] and references therein. In [45, 111] a theory of the function field gamma function is also developed.

The results regarding K-binomial coefficients are taken from [68].

Our exposition of overconvergence phenomena follows [72].

5
The Carlitz rings

In this chapter we consider the Carlitz rings, the rings of differential operators with Carlitz derivatives. After a description of basic algebraic properties of these rings, we study modules over them, introduce a class of quasiholonomic modules, similar in many respects to the class of holonomic modules of the characteristic zero theory. We show that some special functions of analysis over a local field of positive characteristic, as well as partial differential operators of evolution type, generate quasiholonomic modules.

5.1 Algebraic preliminaries

Here we collect some well-known results about noncommutative rings and modules over them, which will be needed in the sequel. The results are given without proofs, which can be found, together with further details, in [30, 79, 10, 12].

5.1.1. Filtered rings. A *filtered ring* is a ring R with a family $\{R_n, \ n = 0, 1, 2, \ldots\}$ of its additive subgroups, such that
(i) for each i, j, $R_i R_j \subset R_{i+j}$;
(ii) for $i < j$, $R_i \subset R_j$;
(iii) $\bigcup_{n=0}^{\infty} R_n = R$.

The family $\{R_n\}$ is called *a filtration* of R.

A *graded ring* is a ring T together with a family $\{T_n, \ n = 0, 1, 2, \ldots\}$ of its additive subgroups, such that
(i) $T_i T_j \subset T_{i+j}$;
(ii) $T = \bigoplus_{n=0}^{\infty} T_n$, as an Abelian group.

The family $\{T_n\}$ is called *a grading* of T; a nonzero element of T_n is said to be *homogeneous* of degree n. If a and b are homogeneous elements and $ab \neq 0$, then $\deg(ab) = \deg a + \deg b$.

If T is a graded ring, then T has a natural filtration $\{R_n\}$ with $R_n = T_0 \oplus \cdots \oplus T_n$. Conversely, given a filtered ring R, one can construct an *associated graded ring* $T = \operatorname{gr} R$ as follows. Set $T_n = R_n/R_{n-1}$, $R_{-1} = \{0\}$, $T = \oplus T_n$. To define multiplication in T, it suffices to consider homogeneous elements. If $a \in R_n \setminus R_{n-1}$, $c \in R_m \setminus R_{m-1}$, we consider $\bar{a} = a + R_{n-1} \in T_n$, $\bar{c} = c + R_{m-1} \in T_m$, and set $\overline{ac} = ac + R_{m+n-1}$. It is easy to check that this multiplication is well-defined and makes T a ring.

Proposition 5.1 *If R is a filtered ring, and $\operatorname{gr} R$ is right (left) Noetherian, then R is right (left) Noetherian.*

Let R be a finitely generated extension of a subring R_0 with generators $x_1, \ldots, x_n$. An element of R of the form $r_0 x_{i_1} r_1 x_{i_2} r_2 \cdots x_{i_l} r_l$ with $r_i \in R_0$ is called a word of length l. The *standard filtration* on R is obtained if R_j ($j \geq 1$) is an additive subgroup of R generated by all words of length $\leq j$.

A finitely generated extension R of a subring R_0 with generators $x_1, \ldots, x_n \notin R_0$ is called an *almost normalizing extension*, if for any i, j
(i) $R_0 x_i + R_0 = x_i R_0 + R_0$;
(ii) $x_i x_j - x_j x_i \in \sum_{l=1}^{n} x_l R_0 + R_0$.

Proposition 5.2 *Let R be an almost normalizing extension of a right (left) Noetherian ring R_0, with the standard filtration. Then both R and $\operatorname{gr} R$ are right (left) Noetherian rings.*

5.1.2. Filtered modules. Let M be a left module over a filtered ring R with a filtration $\{R_n\}$. A *filtration* in M is a system of subgroups M_j, $j = 0, 1, 2, \ldots$, such that
(i) for each i, j, $R_j M_i \subset M_{i+j}$ (as usual, we denote by $R_j M_i$ the set of all finite sums $\sum_l r_l m_l$, $r_l \in R_j$, $m_j \in M_i$);
(ii) for each $i < j$, $M_i \subset M_j$;
(iii) $\bigcup_{n=0}^{\infty} M_n = M$.

A module with a filtration is called a *filtered module*.

The filtration of a module is not unique. In particular, given any submodule M_0 of M which generates M, one can define a *standard* filtration setting $M_n = R_n M_0$.

Let $T = \bigoplus_{n=0}^{\infty} T_n$ be a graded ring. A *grading* of a left T-module M is an Abelian group decomposition $M = \bigoplus_{n=0}^{\infty} M_n$, such that $T_j M_i \subset M_{i+j}$ for any i,j. A *graded module* is a module together with a fixed grading.

The nonzero elements of the subgroup M_n are called homogeneous of degree n, and if $m = \sum_{n=0}^{\infty} m_n$ with $m_n \in M_n$, then m_n is the n-th homogeneous component of m. A graded submodule N of M is a submodule with a grading $N = \bigoplus_{n=0}^{\infty} N_n$, such that $N_n \subset M_n$. A graded homomorphism θ is a T-homomorphism $M \to M'$ between two graded T-modules, such that $\theta(M_n) \subset M'_n$ for each n.

Similarly, if M, M' are filtered R-modules, with the filtrations $\{M_n\}$, $\{M'_n\}$, then a filtered homomorphism θ is a R-homomorphism $\theta : M \to M'$ such that $\theta(M_n) \subset M'_n$ for each n.

For a filtered module M over a filtered ring R, we can define a graded module $\operatorname{gr} M$ over the graded ring $\operatorname{gr} R$, as follows. Let $(\operatorname{gr} M)_n = M_n/M_{n-1}$ ($M_{-1} = \{0\}$), $\operatorname{gr} M = \bigoplus_{n=0}^{\infty} (\operatorname{gr} M)_n$. The multiplication is defined, via the distributive law, by the multiplication of representatives of the residue classes.

Classes of filtered modules over a given filtered ring and graded modules over a given graded ring, with appropriate classes of homomorphisms, form categories. Then the correspondence $M \mapsto \operatorname{gr} M$ is a functor between these categories.

If M, N are filtered R-modules, $\theta : M \to N$ is a filtered homomorphism, then $\theta(M_j) \subset \theta(M) \cap N_j$. If $\theta(M_j) = \theta(M) \cap N_j$ for each j, then θ is called *strict*.

Proposition 5.3 *Let R be a filtered ring, and let $L \xrightarrow{\theta} M \xrightarrow{\varphi} N$ be an exact sequence of filtered modules and filtered homomorphisms. Then $\operatorname{gr} L \xrightarrow{\operatorname{gr} \theta} \operatorname{gr} M \xrightarrow{\operatorname{gr} \varphi} \operatorname{gr} N$ is an exact sequence if and only if θ and φ are strict.*

Corollary 5.4 *Let R be a filtered ring, and $\varphi : M \to N$ a filtered* homomorphism. Then $\operatorname{gr}\varphi$ is injective (surjective) if and only if φ is injective (resp. surjective) and φ is strict.

A filtration $\{M_n\}$ in a module M over a filtered ring is called *good* if the graded module $\operatorname{gr} M$ is finitely generated.

Proposition 5.5 (i) *If a filtered module M has a good filtration, then M is finitely generated.*

(ii) *A standard filtration in a finitely generated module over a filtered ring is good.*

Let M be a filtered R-module, and N be a submodule in M. Let us construct and study filtrations in N and M/N.

A filtration in N can be defined as $N_k = N \cap M_k$. In M/N, we consider the subgroups $F_k = M_k/(M_k \cap N)$, the images of M_k in M/N. The sequence $\{F_k,\ k=0,1,2,\ldots\}$ is a filtration in M/N, and
$$F_k/F_{k-1} \cong M_k/(M_{k-1} + M_k \cap N).$$
This defines the canonical projection
$$\pi_k:\ M_k/M_{k-1} \longrightarrow F_k/F_{k-1}.$$
On the other hand, the imbedding $N \subset M$ generates a monomorphism
$$\varphi_k:\ (N \cap M_k)/(N \cap M_{k-1}) \longrightarrow M_k/M_{k-1}.$$
Passing to the maps of the associated graded modules and using Proposition 5.3 we obtain the exact sequence
$$0 \longrightarrow \operatorname{gr} N \xrightarrow{\varphi} \operatorname{gr} M \xrightarrow{\pi} \operatorname{gr} M/N \longrightarrow 0 \tag{5.1}$$
where $\varphi = \oplus \varphi_k$, $\pi = \oplus \pi_k$.

Suppose that R is a Noetherian ring and M is a filtered R-module with a good filtration. Then $\operatorname{gr} M$ is a finitely generated module over the ring $\operatorname{gr} R$, which is Noetherian by Proposition 5.1. Therefore $\operatorname{gr} M$ is a Noetherian module (see [75]). Due to the exact sequence (5.1), $\operatorname{gr} N$ is isomorphic to a submodule of $\operatorname{gr} M$, while $\operatorname{gr} M/N$ is isomorphic to a quotient module of $\operatorname{gr} M$. Thus, the modules $\operatorname{gr} N$ and $\operatorname{gr} M/N$ are Noetherian, so that the above filtrations in N and M/N (called the *induced filtrations*) are good.

5.1.3. Generalized Weyl algebras (GWA). We will consider only a

GWA of degree 1 [10, 11, 12]. Let D be a ring with an automorphism σ and a central element a. A GWA $A = D(\sigma, a)$ is the ring generated by D and two indeterminates X, Y, with the relations

$$X\lambda = \sigma(\lambda)X, \ Y\lambda = \sigma^{-1}(\lambda)Y \text{ (for all } \lambda \in D), \ YX = a, \ XY = \sigma(a).$$

We have $A = \bigoplus_{n=-\infty}^{\infty} A_n$ where $A_n = Dv_n$,

$$v_n = \begin{cases} X^n, & \text{if } n > 0; \\ 1, & \text{if } n = 0; \\ Y^{-n}, & \text{if } n < 0. \end{cases}$$

It follows from the above relations that

$$v_n v_m = (n, m) v_{n+m} = v_{n+m} \langle n, m \rangle$$

for some $(n, m) = \sigma^{-n-m}(\langle n, m \rangle) \in D$. If $n > 0$ and $m > 0$, then for $n \geq m$

$$(n, -m) = \sigma^n(a) \cdots \sigma^{n-m+1}(a), \quad (-n, m) = \sigma^{-n+1}(a) \cdots \sigma^{-n+m}(a).$$

For $n \leq m$,

$$(n, -m) = \sigma^n(a) \cdots \sigma(a), \quad (-n, m) = \sigma^{-n+1}(a) \cdots a.$$

In other cases $(n, m) = 1$.

If D is a Noetherian domain, then so is A too.

The first and basic example of a GWA was the first Weyl algebra of ordinary differential operators (in a variable ξ) with polynomial coefficients over $\mathbb{C}$. In this case $D = \mathbb{C}[H]$, $\sigma: H \mapsto H - 1$, $a = H$, X is the operator of multiplication by ξ, $Y = d/d\xi$, $H = YX$. Below, the ring of differential operators with Carlitz derivatives (in one variable) will be interpreted as a GWA.

Suppose that the ring D is commutative. The cyclic group G, generated by the automorphism σ, acts on the set Specm(D) of maximal ideals of D. An orbit $\mathfrak{O}$ is *cyclic* of length n (respectively, *linear*) if it contains a finite (respectively, infinite) number $n = $ card $\mathfrak{O}$ of elements. The set of all cyclic (linear) orbits is denoted by Cyc (resp. Lin).

An A-module M (below we consider left modules) is called a *weight module* if V is semisimple as a D-module, so that

$$M = \bigoplus_{\mathfrak{p} \in \text{Specm}(D)} M_{\mathfrak{p}}$$

where $M_{\mathbf{p}}$ (the component of M of weght $\mathbf{p}$) is the sum of simple D-submodules isomorphic to $D/\mathbf{p}$. The support $\mathrm{Supp}(M)$ of the weight module M is the set of maximal ideals $\mathbf{p}$, such that $M_{\mathbf{p}} \neq \{0\}$.

Each weight A-module M is decomposed into the direct sum of A-submodules

$$M = \bigoplus \{M_{\mathfrak{O}} : \mathfrak{O} \text{ is an orbit}\},$$

where $M_{\mathfrak{O}} = \bigoplus \{M_{\mathbf{p}} : \mathbf{p} \in \mathfrak{O}\}$. Hence, the support of a simple weight module belongs exactly to one orbit. Now the set $\widehat{A}(\text{weight})$ of isomorphism classes of simple weight A-modules consists of the sets $\widehat{A}(\text{weight, linear})$ and $\widehat{A}(\text{weight, cyclic})$ of isomorphism classes of simple weight A-modules with support from a linear resp. cyclic orbit.

An orbit $\mathfrak{O}$ is called *degenerate* if it contains an ideal $\mathbf{p}$ (a *marked* ideal) such that $a \in \mathbf{p}$. Denote by Linn (Cycn) the set of all nondegenerate linear (cyclic) orbits.

Each linear orbit $\mathfrak{O}(\mathbf{p})$, $\mathbf{p} \in \mathfrak{O}$ may be identified with $\mathbb{Z}$ via the mapping $\sigma^i(\mathbf{p}) \mapsto i$. Therefore the notation used for $\mathbb{Z}$ (like an interval, semiaxis, etc) may be used for $\mathfrak{O}(\mathbf{p})$. For example, $\sigma^i(\mathbf{p}) \leq \sigma^j(\mathbf{p})$ if and only if $i \leq j$; $(-\infty, \mathbf{p}] = \{\sigma^i(\mathbf{p}), i \leq 0\}$. Marked ideals $\mathbf{p}_1 < \ldots < \mathbf{p}_s$ of a degenerate linear orbit $\mathfrak{O}$ divide it into $s+1$ parts,

$$\Gamma_1 = (-\infty, \mathbf{p}_1], \ \Gamma_2 = (\mathbf{p}_1, \mathbf{p}_2], \ldots, \Gamma_{s+1} = (\mathbf{p}_s, \infty).$$

For maximal ideals belonging to linear orbits (we write $\mathbf{p} \in \mathrm{Specm.\,lin}(D)$), we introduce the equivalence relation: $\mathbf{p} \sim \mathbf{q}$ if and only if $\mathbf{p}, \mathbf{q} \in \mathrm{Supp}\, M$ for some isomorphism class $[M] \in \widehat{A}(\text{weight, linear})$. It can be proved that $\mathbf{p} \sim \mathbf{q}$ if and only if $\mathbf{p}$ and $\mathbf{q}$ both belong either to a nondegenerate linear orbit or to some Γ_i.

Theorem 5.6 *The mapping*

$$\mathrm{Specm.\,lin}(D)/\sim \ \longrightarrow \ \widehat{A}(\text{weight, linear}), \quad \Gamma \mapsto [L(\Gamma)],$$

is bijective, with the inverse $[L] \mapsto \mathrm{Supp}\, L$. *Here*
(i) *if* $\Gamma \in Linn$ *is a nondegenerate orbit, then* $L(\Gamma) = A/A\mathbf{p}$, $\mathbf{p} \in \Gamma$;
(ii) *if* $\Gamma = (-\infty, \mathbf{p}]$, *then* $L(\Gamma) = A/A(\mathbf{p}, X)$;
(iii) *if* $\Gamma = (\sigma^{-n}(\mathbf{p}), \mathbf{p}]$, $n \in \mathbb{N}$, *then* $L(\Gamma) = A/A(\mathbf{p}, X, Y^n)$;
(iv) *if* $\Gamma = (\mathbf{p}, \infty)$, *then* $L(\Gamma) = A/A(\sigma(\mathbf{p}), Y)$.

See [12] for a description of $\widehat{A}(\text{weight, cyclic})$.

Let D be a Dedekind domain (for example, a principal ideal domain [6]), M be an A-module. Let $\mathrm{tor}(M) = \{m \in M : am = 0 \text{ for some } 0 \neq a \in D\}$ be the D-torsion submodule of M. Every simple A-module is either D-torsion ($M = \mathrm{tor}(M)$) or D-torsion-free ($\mathrm{tor}(M) = \{0\}$). Moreover, M is a weight module if and only if it is D-torsion.

Theorem 5.7 *All simple A-modules are D-torsion if and only if there are inifinitely many cyclic orbits.*

5.2 The Carlitz rings

5.2.1. Definitions and general properties.

As in Chapter 3 (see (3.57)) we denote by $\mathcal{F}_{n+1}$ the set of all germs of functions of the form

$$f(s, t_1, \ldots, t_n) = \sum_{k_1=0}^{\infty} \cdots \sum_{k_n=0}^{\infty} \sum_{m=0}^{\min(k_1,\ldots,k_n)} a_{m,k_1,\ldots,k_n} s^{q^m} t_1^{q^{k_1}} \ldots t_n^{q^{k_n}} \quad (5.2)$$

where $a_{m,k_1,\ldots,k_n} \in \overline{K}_c$ are such that all the series are convergent on some neighborhoods of the origin. We do not exclude the case $n = 0$ where $\mathcal{F}_1$ will mean the set of all $\mathbb{F}_q$-linear power series $\sum_m a_m s^{q^m}$ convergent on a neighborhood of the origin. $\widehat{\mathcal{F}}_{n+1}$ will denote the set of all polynomials from $\mathcal{F}_{n+1}$, that is the series (5.2) in which only a finite number of coefficients is different from zero.

The ring $\mathfrak{A}_{n+1}$ is generated by the operators $\tau, d = d_s, \Delta_{t_1}, \ldots, \Delta_{t_n}$ on $\mathcal{F}_{n+1}$, and the operators of multiplication by scalars from $\overline{K}_c$. To simplify the notation, we will write Δ_j instead of Δ_{t_j} and identify a scalar $\lambda \in \overline{K}_c$ with the operator of multiplication by λ. The operators Δ_j are $\overline{K}_c$-linear, so that

$$\Delta_j \lambda = \lambda \Delta_j, \quad \lambda \in \overline{K}_c, \quad (5.3)$$

while the operators τ, d_s satisfy the commutation relations

$$d\tau - \tau d = [1]^{1/q}, \ \tau\lambda = \lambda^q \tau, \ d\lambda = \lambda^{1/q} d \ (\lambda \in \overline{K}_c). \quad (5.4)$$

In the action of each operator d_s, Δ_j (acting in a single variable), other variables are treated as scalars. The operator τ acts simultaneously on all the variables and coefficients, so that

$$\tau f = \sum a_{m,k_1,\ldots,k_n}^q s^{q^{m+1}} t_1^{q^{k_1+1}} \ldots t_n^{q^{k_n+1}}.$$

$\mathfrak{A}_{n+1}$ is a $\mathbb{F}_q$-algebra, but is not a K-algebra.

We have
$$\Delta_j t_j^{q^k} = \begin{cases} [k]t_j^{q^k}, & \text{if } k \geq 1; \\ 0, & \text{if } k = 0; \end{cases} \tag{5.5}$$

the second equality can be included in the first one, if we set $[0] = 0$. Similarly
$$d_s s^{q^m} = [m]^{1/q} s^{q^{m-1}}, \quad m \geq 0. \tag{5.6}$$

Since $|[m]| = q^{-1}$ for any $m \geq 1$, the action of operators from $\mathfrak{A}_{n+1}$ does not spoil convergence of the series (5.2).

The identity $[k+1] - [k]^q = [1]$, together with (5.5) and (5.6), implies the commutation relations
$$\Delta_j \tau - \tau \Delta_j = [1]\tau, \quad d_s \Delta_j - \Delta_j d_s = [1]^{1/q} d_s, \quad j = 1, \ldots, n, \tag{5.7}$$

verified by applying both sides of each equality to an arbitrary monomial.

As we saw in Section 3.4.1, operators from $\mathfrak{A}_{n+1}$ possess properties resembling those of a class of partial differential operators of classical analysis. In many respects, the rings $\mathfrak{A}_{n+1}$ may be seen as the function field counterparts of the Weyl algebras of polynomial differential operators [16, 30].

Using the commutation relations (5.3), (5.4), and (5.7), we can write any element $a \in \mathfrak{A}_{n+1}$ as a finite sum
$$a = \sum c_{l,\mu,i_1,\ldots,i_n} \tau^l d_s^\mu \Delta_1^{i_1} \ldots \Delta_n^{i_n}. \tag{5.8}$$

Proposition 5.8 *The representation (5.8) of an element $a \in \mathfrak{A}_{n+1}$ is unique.*

Proof. Suppose that
$$\sum_{l,\mu,i_1,\ldots,i_n} c_{l,\mu,i_1,\ldots,i_n} \tau^l d_s^\mu \Delta_1^{i_1} \ldots \Delta_n^{i_n} = 0. \tag{5.9}$$

Applying the left-hand side of (5.9) to the function $s t_1^{q^{k_1}} \ldots t_n^{q^{k_n}}$ with $k_1, \ldots, k_n > 0$ we find that
$$\sum_l \left(\sum_{i_1,\ldots,i_n} c_{l,0,i_1,\ldots,i_n} [k_1]^{i_1 q^l} \ldots [k_n]^{i_n q^l} \right) s^{q^l} t_1^{q^{k_1+l}} \ldots t_n^{q^{k_n+l}} = 0$$

whence
$$\sum_{i_1,\ldots,i_n} c_{l,0,i_1,\ldots,i_n}[k_1]^{i_1 q^l}\ldots[k_n]^{i_n q^l} = 0$$

for each l. Writing this in the form
$$\sum_{i_n} \rho(i_n) y^{i_n} = 0 \qquad (5.10)$$

where
$$\rho(i_n) = \sum_{i_1,\ldots,i_{n-1}} c_{l,0,i_1,\ldots,i_n}[k_1]^{i_1 q^l}\ldots[k_{n-1}]^{i_{n-1} q^l}, \quad y = [k_n]^{q^l},$$

and taking into account that (5.10) holds for arbitrary $k_n \geq 1$, that is for an infinite set of values of y, we find that $\rho(i_n) = 0$. Repeating this reasoning we get the equality $c_{l,0,i_1,\ldots,i_n} = 0$ for all $l, 0, i_1, \ldots, i_n$.

Suppose that $c_{l,\mu,i_1,\ldots,i_n} = 0$ for $\mu \leq \mu_0$ and arbitrary $l, i_1, \ldots, i_n$. Then we apply the left-hand side of (5.9) to the function $s^{q^{\mu_0+1}} t_1^{q^{k_1}} \ldots t_n^{q^{k_n}}$ and proceed as before coming to the equality $c_{l,\mu_0+1,i_1,\ldots,i_n} = 0$ for all $l, i_1, \ldots, i_n$. ∎

The basic algebraic property of the ring $\mathfrak{A}_{n+1}$ is given by the following result.

Theorem 5.9 $\mathfrak{A}_{n+1}$ *is a Noetherian domain.*

Proof. It is clear that $\mathfrak{A}_{n+1}$ is an almost normalizing extension of the field $\overline{K}_c$, so that, by Proposition 5.2, it is a Noetherian ring.

Let us prove that $\mathfrak{A}_{n+1}$ has no zero-divisors. We begin with the case $n = 0$. For brevity, we write d instead of d_s. Let $a, b \in \mathfrak{A}_1$, $ab = 0$,
$$a = \sum_{i=0}^{m_1}\sum_{j=0}^{n_1} \lambda_{ij}\tau^i d^j, \quad b = \sum_{k=0}^{m_2}\sum_{l=0}^{n_2} \mu_{kl}\tau^k d^l,$$

and $a \neq 0$, $b \neq 0$, that is
$$\sum_{i=0}^{m_1} \lambda_{i n_1}\tau^i \neq 0, \quad \sum_{k=0}^{m_2} \mu_{k n_2}\tau^k \neq 0. \qquad (5.11)$$

It is easily proved (by induction) that
$$d\tau^i - \tau^i d = [i]^{1/q}\tau^{i-1}, \quad d^j\tau - \tau d^j = [j]^{q^{-j}} d^{j-1}, \qquad (5.12)$$

for any natural numbers i, j. It follows from (5.12) that
$$d^j \tau^k = \tau^k d^j + O(d^{j-1})$$
where $O(d^{j-1})$ means a polynomial in the variable d, of degree $\leq j-1$, with coefficients from the composition ring $\overline{K}_c\{\tau\}$ of polynomials in the operator τ.

Therefore the coefficient of $d^{n_1+n_2}$ in the expression for the operator ab equals
$$\sum_{i,k} \lambda_{in_1} \mu_{kn_2}^{i-n_1} \tau^{i+k} = P_1(P_2(\tau))$$
where
$$P_1(\tau) = \sum_{i=0}^{m_1} \lambda_{in_1} \tau^i, \quad P_2(\tau) = \sum_{k=0}^{m_2} \mu_{kn_2}^{q^{-n_1}} \tau^k,$$
which contradicts (5.11), since the ring $\overline{K}_c\{\tau\}$ has no zero-divisors [45].

Suppose that it has been proved that $\mathfrak{A}_{n+1}$ has no zero-divisors for some n. Let us prove that $\mathfrak{A}_{n+2}$ has no zero-divisors.

Let $a, b \in \mathfrak{A}_{n+2}$, $a \neq 0$, $b \neq 0$, and $ab = 0$. Let us write the representations (5.8) for the elements a and b in the form of (noncommutative) polynomials
$$a = \sum_{i=0}^{k} a_i \Delta_{n+1}^i, \quad b = \sum_{j=0}^{l} b_j \Delta_{n+1}^j,$$
with coefficients from $\mathfrak{A}_{n+1}$, such that $a_k \neq 0$, $b_l \neq 0$.

It follows from the commutation relations (5.3) and (5.7) that the degrees of polynomials in Δ_{n+1} obtained after multiplication of monomials $a_i \Delta_{n+1}^i$ and $b_j \Delta_{n+1}^j$ cannot exceed $i+j$. In particular,
$$\Delta_{n+1}^k b_l = b_l \Delta_{n+1}^k + O\left(\Delta_{n+1}^{k-1}\right)$$
(here the symbol O means a polynomial with coefficients from $\mathfrak{A}_{n+1}$), so that
$$a_k \Delta_{n+1}^k b_l \Delta_{n+1}^l = a_k b_l \Delta_{n+1}^{k+l} + O\left(\Delta_{n+1}^{k+l-1}\right).$$
Now the assumption $ab = 0$ means, since $\mathfrak{A}_{n+1}$ has no zero-divisors, that $a_k = 0$ or $b_l = 0$, and we come to a contradiction. ∎

5.2.2. Filtration in $\mathfrak{A}_{n+1}$. Let us introduce a filtration in $\mathfrak{A}_{n+1}$ denoting by Γ_ν, $\nu \in \mathbb{Z}_+$, the $\overline{K}_c$-vector space of operators (5.8) with max$\{l + \mu + i_1 + \cdots + i_n\} \leq \nu$ where the maximum is taken over all the terms contained in the representation (5.8). It is clear that $\mathfrak{A}_{n+1}$ is a filtered ring. Setting $T_0 = \overline{K}_c$, $T_\nu = \Gamma_\nu/\Gamma_{\nu-1}$, $\nu \geq 1$, we introduce the associated graded ring

$$\text{gr}(\mathfrak{A}_{n+1}) = \bigoplus_{\nu=0}^{\infty} T_\nu.$$

It is generated by scalars $\lambda \in T_0$ and the images $\bar{\tau}, \bar{d}_s, \bar{\Delta}_1, \ldots, \bar{\Delta}_n \in T_1$ of the elements $\tau, d_s, \Delta_1, \ldots, \Delta_n \in \Gamma_1$ respectively, which satisfy, by virtue of (5.3), (5.4), and (5.7), the relations

$$\bar{d}_s\bar{\tau} - \bar{\tau}\bar{d}_s = 0, \bar{\tau}\lambda = \lambda^q\bar{\tau}, \bar{d}_s\lambda = \lambda^{1/q}\bar{d}_s,$$
$$\bar{d}_s\bar{\Delta}_j - \bar{\Delta}_j\bar{d}_s = 0, \bar{\Delta}_j\bar{\tau} - \bar{\tau}\bar{\Delta}_j = 0, \bar{\Delta}_j\lambda = \lambda\bar{\Delta}_j \quad (j = 1, \ldots, n).$$

By Proposition 5.2, $\text{gr}(\mathfrak{A}_{n+1})$ is Noetherian.

Let us compute the dimension of the $\overline{K}_c$-vector space Γ_ν. Note that

$$\dim \Gamma_\nu = \dim \bigoplus_{j=1}^{\nu} T_j,$$

so that $\dim \Gamma_\nu$ coincides with the dimension of the appropriate space appearing in the natural filtration in $\text{gr}(\mathfrak{A}_{n+1})$.

Lemma 5.10 *For any $\nu \in \mathbb{N}$,*

$$\dim \Gamma_\nu = \binom{\nu + n + 2}{n + 2}.$$

Proof. The number $\dim \Gamma_\nu$ coincides with the number of nonnegative integral solutions $(l, \mu, i_1, \ldots, i_n)$ of the inequality $l + \mu + i_1 + \cdots + i_n \leq \nu$, so that

$$\dim \Gamma_\nu = \sum_{j=0}^{\nu} N(j, n+2)$$

where $N(j, k)$ is the number of different representations of j as sums of k nonnegative integers. It is known (Proposition 6.1 in [74]) that $N(j, k) = \binom{j + k - 1}{k - 1}$. Then (see Section 1.3 from [88])

$$\dim \Gamma_\nu = \sum_{j=0}^{\nu} \binom{j + n + 1}{n + 1} = \sum_{i=0}^{\nu} \binom{\nu + n + 1 - i}{n + 1} = \binom{\nu + n + 2}{n + 2},$$

as desired. ∎

5.3 The ring $\mathfrak{A}_1$

5.3.1. $\mathfrak{A}_1$ as a GWA. Let us consider a GWA $A = D(\sigma, H)$ where $D = \overline{K}_c[H]$, σ is the $\mathbb{F}_q$-linear automorphism of $\overline{K}_c[H]$ defined by the assignments:

$$\sigma: H \mapsto H - \lambda_1, \ \lambda \mapsto \lambda^q \ (\lambda \in \overline{K}_c),$$

$\lambda_1 = [1]^{1/q} = x - x^{1/q}$; see Section 5.1.3 for the notation related to a GWA.

It is easy to check that the mapping

$$\mathfrak{A}_1 \to D(\sigma, H), \ \tau \mapsto X, d \mapsto Y, d\tau \mapsto H, \lambda \mapsto \lambda \ (\lambda \in \overline{K}_c),$$

is an $\mathbb{F}_q$-algebra isomorphism. Thus $\mathfrak{A}_1$ can be identified with the above GWA.

Let $G = \langle \sigma \rangle$ be the subgroup of the group of ring automorphisms $\mathrm{Aut}(D)$ of D generated by the element σ. The group G acts in the obvious way on the set of maximal ideals of the algebra D, $\mathrm{Specm}(D) = \{D(H - \lambda) | \ \lambda \in \overline{K}_c\} \cong \overline{K}_c$, $D(H - \lambda) \leftrightarrow \lambda$. The orbit $\mathfrak{O}$ of an element $\mathbf{p} = D(H - \lambda) \in \mathrm{Specm}(D)$ for some $\lambda \in \overline{K}_c$ is determined by the elements

$$\sigma^n(H - \lambda) = H - \lambda_1 - \lambda_1^q - \cdots - \lambda_1^{q^{n-1}} - \lambda^{q^n}, \quad (5.13)$$

$$\sigma^{-n}(H - \lambda) = H + \lambda_1 + \lambda_1^{\frac{1}{q}} + \cdots + \lambda_1^{\frac{1}{q^{n-1}}} - \lambda^{\frac{1}{q^n}}, \quad (5.14)$$

$n \geq 1$. Note that, for $n \geq 2$,

$$\lambda_1 + \lambda_1^q + \cdots + \lambda_1^{q^n} = -x^{1/q} + x^{q^n},$$

$$\lambda_1 + \lambda_1^{\frac{1}{q}} + \cdots + \lambda_1^{\frac{1}{q^n}} = -x^{\frac{1}{q^{n+1}}} + x.$$

Lemma 5.11 *Let $\lambda \in \overline{K}_c$, $n \in \mathbb{N}$. Then $\sigma^n(D(H - \lambda)) = D(H - \lambda)$ if and only if $\lambda \in -x^{1/q} + \mathbb{F}_{q^n}$.*

Proof. If $n = 1$, then $\sigma(D(H - \lambda)) = D(H - \lambda)$ if and only if $g_1(\lambda) = 0$ where $g_1(\lambda) = \lambda^q - \lambda + \lambda_1 = \lambda^q - \lambda - x^{1/q} + x$. The polynomial g_1 of degree q has q distinct roots, since its derivative equals -1. Since $g_1(\lambda + \gamma) = g_1(\lambda) + \gamma^q - \gamma$, we find that for a root λ of g_1, $\lambda + \gamma$ is a root for any $\gamma \in \mathbb{F}_q$.

Noticing that $-x^{1/q}$ is a root we obtain that $-x^{1/q} + \mathbb{F}_q$ is the set of all roots.

Similarly, if $n \geq 2$ then, by (5.13), $\sigma^n(D(H - \lambda)) = D(H - \lambda)$ if and only if $g_n(\lambda) = 0$,

$$g_n(\lambda) = \lambda^{q^n} - \lambda + \lambda_1 + \lambda_1^q + \cdots + \lambda_1^{q^{n-1}} = \lambda^{q^n} - \lambda - x^{1/q} + x^{q^{n-1}}.$$

As above, we find that $-x^{1/q} + \mathbb{F}_{q^n}$ are the roots of g_n. ∎

Below we will need the *Möbius function* $\mu : \mathbb{N} \to \{0, \pm 1\}$ given by the rule: $\mu(1) = 1$, $\mu(p_1 \cdots p_r) = (-1)^r$ if $p_1, \ldots, p_r$ are distinct primes, and $\mu(n) = 0$ otherwise. The Möbius inversion formula (Theorem 3.24 in [76]) is as follows. Let f be a function on $\mathbb{N}$ with values in an Abelian group, and $g(n) = \sum_{d|n} f(d)$. Then

$$f(n) = \sum_{d|n} \mu(d) g\left(\frac{n}{d}\right). \tag{5.15}$$

The *Euler function* φ on $\mathbb{N}$ is defined as $\varphi(1) = 1$ and $\varphi(n)$ being the number of natural numbers m that are coprime to n and $1 \leq m < n$.

The next result gives a classification of finite orbits in $\mathrm{Specm}(D)$.

Theorem 5.12 *Let $\mathfrak{O} = \mathfrak{O}_\lambda$ ($\lambda \in \overline{K}_c$) be the orbit of a maximal ideal $D(H - \lambda)$ of the polynomial algebra $D = \overline{K}_c[H]$ under the action of the cyclic group G. Then:*

(i) The orbit $\mathfrak{O}_\lambda$ consists of a single element if and only if $\lambda \in -x^{1/q} + \mathbb{F}_q$. Thus there are exactly q distinct maximal σ-invariant ideals of the algebra D.

(ii) The orbit $\mathfrak{O}_\lambda$ contains exactly $n \geq 2$ elements, if and only if

$$\lambda \in -x^{1/q} + \left(\mathbb{F}_{q^n} \setminus \bigcup_{m|n, m \neq n} \mathbb{F}_{q^m}\right).$$

Thus there are exactly

$$l_n = n^{-1} \sum_{d|n} \mu(d) q^{\frac{n}{d}} = n^{-1} \varphi(q^n - 1)$$

distinct orbits, such that each of them contains exactly $n \geq 2$ elements, and $l_n \geq n^{-1} q^n \left(1 - \frac{1}{q-1}\right) > 0.$

Proof. The statement (i) follows immediately from Lemma 5.11. By the same lemma, the orbit $\mathfrak{O}_\lambda$ contains exactly $n \geq 2$ elements if and only if

$$\lambda \in -x^{1/q} + \left(\mathbb{F}_{q^n} \setminus \bigcup_{m|n, m \neq n} \mathbb{F}_{q^m} \right).$$

Let l_n be the number of distinct orbits containing exactly n elements. Then $f(n) = nl_n$ is the number of all maximal ideals of D belonging to the above orbits. It follows from Lemma 5.11 that the function $g(n) = \sum_{d|n} f(d)$ is equal to $\mathrm{card}\left(-x^{1/q} + \mathbb{F}_{q^n}\right) = \mathrm{card}\left(\mathbb{F}_{q^n}\right) = q^n$. Therefore, by (5.15), we have

$$f(n) = \sum_{d|n} \mu(d) q^{\frac{n}{d}},$$

whence $l_n = n^{-1} \sum_{d|n} \mu(d) q^{\frac{n}{d}}$. Next,

$$nl_n \geq \mathrm{card}\left(-x^{1/q} + \mathbb{F}_{q^n}\right) - \mathrm{card}\left(-x^{1/q} + \mathbb{F}_{q^{n-1}}\right)$$
$$- \cdots - \mathrm{card}\left(-x^{1/q} + \mathbb{F}_q\right) = q^n - q^{n-1} - \cdots - q$$
$$= q^n - q \frac{q^{n-1} - 1}{q - 1} > q^n - \frac{q^n}{q-1} = q^n \left(1 - \frac{1}{q-1} \right) > 0.$$

Clearly, $l_n = n^{-1} \mathrm{card}\left(\mathbb{F}_{q^n} \setminus \bigcup_{m|n, m \neq n} \mathbb{F}_{q^m} \right) = n^{-1} \varphi(q^n - 1)$, because $\varphi(q^n - 1)$ equals the number of primitive elements of $\mathbb{F}_{q^n}$ [76]. ∎

5.3.2. Simple $\mathfrak{A}_1$-modules. The above description of cyclic orbits, together with the general results of Section 5.1.3, leads to a classification of simple $\mathfrak{A}_1$-modules. First of all, note that, by Theorems 5.7 and 5.12, *all simple $\mathfrak{A}_1$-modules are weight.*

Denote by $\widehat{\mathfrak{A}}_1(\text{linear})$ and $\widehat{\mathfrak{A}}_1(\text{cyclic})$ the sets of isomorphism classes of simple $\mathfrak{A}_1$-modules with support from a linear and a cyclic orbit respectively.

The ideal (H) of the polynomial algebra $D = \overline{K}_c[H]$ generated by the element H is a maximal ideal of D. By (5.13), we see that the orbit $\mathfrak{O}(H)$ of the maximal ideal (H) is an infinite orbit (since $\lambda_1^{q^n} \neq \lambda_1$ for all $n \geq 1$). This linear orbit is the only degenerate one. Therefore, in terms of the equivalence introduced in Section 5.1.3, there are only two equivalence classes in $\mathfrak{O}(H)$: $\Gamma_- = (-\infty, (H)]$ and $\Gamma_+ = ((H), \infty)$. Correspondingly,

The Carlitz rings

the set of equivalence classes in Specm. lin(D) consists of Γ_-, Γ_+, and the nondegenerate linear orbits.

Now Theorem 5.6 specializes to the following description.

Theorem 5.13 *The mapping*

$$\text{Specm. lin}(D)/\sim \longrightarrow \widehat{\mathfrak{A}}_1(\text{linear}), \quad \Gamma \mapsto [L(\Gamma)],$$

is a bijection, with inverse $[L] \mapsto \text{Supp } L$ *(in particular,* $\text{Supp } L(\Gamma) = \Gamma$*) where*

(i) *if* $\Gamma \in Linn$, *then* $L(\Gamma) = \mathfrak{A}_1/\mathfrak{A}_1\mathbf{p}$, *for any* $\mathbf{p} \in \Gamma$;
(ii) *if* $\Gamma = \Gamma_- = (-\infty, (H)]$, *then* $L(\Gamma_-) = \mathfrak{A}_1/(\mathfrak{A}_1 H + \mathfrak{A}_1 X)$;
(iii) *if* $\Gamma = \Gamma_+ = ((H), \infty)$, *then* $L(\Gamma_+) = \mathfrak{A}_1/(\mathfrak{A}_1 \sigma(H) + \mathfrak{A}_1 Y)$.

All the modules listed in Theorem 5.13 are infinite-dimensional as vector spaces over $\overline{K}_c$. They can be written in an explicit form as follows.

(i) $L = \mathfrak{A}_1/\mathfrak{A}_1\mathbf{p} = \left(\bigoplus_{i\geq 1} \overline{K}_c \overline{Y}^i\right) \oplus \overline{K}_c \overline{1} \oplus \left(\bigoplus_{i\geq 1} \overline{K}_c \overline{X}^i\right)$

where $\overline{u} = u + \mathfrak{A}_1 \mathbf{p}$, $\overline{1} = \overline{X}^0 = \overline{Y}^0$, $\mathbf{p} = (H-\lambda) \in \Gamma \in Linn$, and the action of the generators of $\mathfrak{A}_1$ (as a GWA) is given by the rules:

$$X\overline{X}^i = \overline{X}^{i+1}, \quad Y\overline{Y}^i = \overline{Y}^{i+1}, \ i \geq 0;$$

$$X\overline{Y}^i = \left(\lambda - \lambda_1 - \lambda_1^q - \cdots - \lambda_1^{q^{i-1}}\right)\overline{Y}^{i-1}, \ i \geq 1;$$

$$Y\overline{X}^i = \left(\lambda + \lambda_1 + \lambda_1^{\frac{1}{q}} + \cdots + \lambda_1^{\frac{1}{q^{i-1}}}\right)\overline{X}^{i-1}, \ i \geq 2;$$

$$Y\overline{X} = \lambda\overline{1}.$$

(ii) $L_- = \bigoplus_{i\geq 0} \overline{K}_c \overline{Y}^i$, $\overline{Y}^i = Y^i + \mathfrak{A}_1 H + \mathfrak{A}_1 X$,

with the action

$$H\overline{Y}^0 = 0, \quad H\overline{Y}^i = -\left(\lambda_1 + \lambda_1^q + \cdots + \lambda_1^{q^{i-1}}\right)\overline{Y}^i, \ i \geq 1;$$

$$X\overline{Y}^0 = 0, \quad X\overline{Y}^i = -\left(\lambda_1 + \lambda_1^q + \cdots + \lambda_1^{q^{i-1}}\right)\overline{Y}^{i-1}, \ i \geq 1;$$

$$Y\overline{Y}^i = \overline{Y}^{i+1}, \quad i \geq 0.$$

(iii) $L_+ = \bigoplus_{i \geq 0} \overline{K}_c \overline{X}^i$, $\overline{X}^i = X^i + \mathfrak{A}_1 \sigma(H) + \mathfrak{A}_1 Y$,

with the action

$$H\overline{X}^0 = \lambda_1 \overline{X}^0, \ H\overline{X}^1 = 2\lambda_1 \overline{X}^1, \ H\overline{X}^i = \left(2\lambda_1 + \lambda_1^{\frac{1}{q}} + \cdots + \lambda_1^{\frac{1}{q^{i-1}}}\right) \overline{X}^i,$$

$i \geq 2$;

$$Y\overline{X}^0 = 0, \ Y\overline{X}^1 = \lambda_1 \overline{X}^0, \ Y\overline{X}^i = \left(2\lambda_1 + \lambda_1^{\frac{1}{q}} + \cdots + \lambda_1^{\frac{1}{q^{i-2}}}\right) \overline{X}^{i-1},$$

$i \geq 2$;

$$X\overline{X}^i = \overline{X}^{i+1}, \quad i \geq 0.$$

For a description of all the simple cyclic $\mathfrak{A}_1$-modules see [11]. It is very important that, in contrast to the case of the Weyl algebra of polynomial differential operators over a field of characteristic zero, there exist finite-dimensional $\mathfrak{A}_1$-modules. A simple construction is given in the proof of the following theorem.

Theorem 5.14 *For any $k = 1, 2, \ldots$, there exists a non-trivial $\mathfrak{A}_1$-module M whose dimension, as of a vector space over $\overline{K}_c$, equals k.*

Proof. Let $M = (\overline{K}_c)^k$. Denote by $\mathbf{e}_1, \ldots, \mathbf{e}_k$ the standard basis in M, that is $\mathbf{e}_j = (0, \ldots, 0, 1, 0, \ldots, 0)$, with 1 at the j-th place. Let (λ_{ij}) be a $k \times k$ matrix over $\overline{K}_c$, such that $\lambda_{ij} \in \mathbb{F}_q$ if $i \neq j$, while the diagonal elements satisfy the equation $\lambda^q - \lambda + [1]^{1/q} = 0$. We define the action of τ and d_s on M as follows:

$$\tau(c\mathbf{e}_j) = c^q \mathbf{e}_j; \ d(\mathbf{e}_j) = \sum_{i=1}^n \lambda_{ij} \mathbf{e}_i; \ d(c\mathbf{e}_j) = c^{1/q} \mathbf{e}_j, \quad c \in \overline{K}_c, j = 1, \ldots, k,$$

with subsequent additive continuation onto M.

If $x = \sum_{j=1}^k c_j \mathbf{e}_j$, $c_j \in \overline{K}_c$, then we have

$$\tau d(x) = \sum_{j=1}^k c_j \sum_{i=1}^n \lambda_{ij}^q \mathbf{e}_i, \quad d\tau(x) = \sum_{j=1}^k c_j \sum_{i=1}^n \lambda_{ij} \mathbf{e}_i,$$

so that
$$d\tau(x) - \tau d(x) = [1]^{1/q}x,$$
and we indeed have an $\mathfrak{A}_1$-module. ∎

5.3.3. The center and ideals of $\mathfrak{A}_1$. Let $\overline{K}_c(H) = S^{-1}D$, $S = D\backslash\{0\}$, be the field of fractions of D. The localization $B = S^{-1}\mathfrak{A}_1$ at S is the skew Laurent polynomial ring $B = \overline{K}_c(H)[X, X^{-1}, \sigma]$, that is (see [79]) the ring of polynomials over $\overline{K}_c(H)$ in X and X^{-1} subject to the relation $aX = X\sigma(a)$, $a \in \overline{K}_c(H)$. The ring $\mathfrak{A}_1$ can be identified with a subring of B via the ring monomorphism
$$\mathfrak{A}_1 \to B, \ X \mapsto X, \ Y \mapsto HX^{-1}, \ \delta \mapsto \delta \ (\delta \in D).$$
The ring B is a Euclidean ring (the left and right division with remainder algorithms hold), hence a principal left and right ideal domain [79].

Returning to the cyclic orbits on $\mathrm{Specm}(D)$, denote by Cyc_n the set of all the finite orbits containing exactly n elements. The quantity of such orbits was counted in Theorem 5.12. For each orbit $\mathfrak{O} \in Cyc_n$ and any λ, such that $D(H - \lambda) \in \mathfrak{O}$, let
$$\alpha_{\mathfrak{O}} = \prod_{i=0}^{n-1} \sigma^i(H - \lambda).$$
It follows from (5.13) that $\alpha_{\mathfrak{O}}$ is a central element of $\mathfrak{A}_1$ (since $\sigma^n(H-\lambda) = H - g_n(\lambda) - \lambda = H - \lambda$; see the proof of Lemma 5.11).

The next result describes the center of the ring $\mathfrak{A}_1$ and its localization B.

Proposition 5.15 (i) *For each $n \in \mathbb{N}$, the skew Laurent extension $\overline{K}_c[t, t^{-1}; \sigma^n]$ is a simple $\mathbb{F}_q$-algebra with center $\mathbb{F}_{q^n}$.*

(ii) *The ring $B = S^{-1}\mathfrak{A}_1 = \overline{K}_c(H)[X, X^{-1}; \sigma]$ is a simple ring with center*
$$Z(B) = \left\{ \mathbb{F}_q \prod_{\mathfrak{O} \in Cyc} \alpha_{\mathfrak{O}}^{n(\mathfrak{O})} : n(\mathfrak{O}) \in \mathbb{Z}, \right.$$
$$\left. \text{and } n(\mathfrak{O}) = 0 \text{ for all but finitely many } \mathfrak{O} \right\}$$
which containes countably many elements.

(iii) *The center*

$$Z(\mathfrak{A}_1) = \left\{ \mathbb{F}_q \prod_{\mathfrak{D} \in Cyc} \alpha_{\mathfrak{D}}^{n(\mathfrak{D})} : n(\mathfrak{D}) \geq 0, \right.$$

$$\left. \text{and all but finitely many } n(\mathfrak{D}) = 0 \right\}$$

contains countably many elements.

Proof. (i) Let Z be the center of $\overline{K}_c[t, t^{-1}; \sigma^n]$. First we prove that $Z \subset \overline{K}_c$. Indeed, otherwise we would have a nonzero central element $z = \lambda t^m + z'$ where $0 \neq m \in \mathbb{Z}$, $0 \neq \lambda \in \overline{K}_c$, and z' contains elements of strictly higher or strictly lower degrees in t. For any $\mu \in \overline{K}_c \setminus \mathbb{F}_{q^{n|m|}}$, we have

$$\mu z - z\mu = (\mu - \sigma^{nm}(\mu)) \lambda t^m + z'' \neq 0$$

(z'' does not contain terms with t^m), and we come to a contradiction. Next, an element $\lambda \in \overline{K}_c$ commutes with t if and only if $\lambda \in \mathbb{F}_{q^n}$. Therefore $Z = \mathbb{F}_{q^n}$.

The fact that $\overline{K}_c[t, t^{-1}; \sigma^n]$ is a simple ring follows from a general result about skew Laurent polynomial rings (Theorem 1.8.5 in [79]).

(ii) Similarly, B is a simple ring. The above reasoning leads to the inclusion $Z(B) \subset \overline{K}_c(H)$. Clearly, $\mathbb{F}_q \prod_{\mathfrak{D} \in Cyc} \alpha_{\mathfrak{D}}^{n(\mathfrak{D})} \subset Z(B)$. We have to prove that any nonzero element $z \in Z(B)$ can be written in this form. The rational function z is equal to $\gamma \frac{f}{g}$ where $f, g \in \mathbb{F}_q[H]$ are coprime monic polynomials, and $\gamma \in \overline{K}_c$.

If $z = \gamma$, then $\gamma = X\gamma X^{-1} = \sigma(\gamma)$, so that $\gamma \in \mathbb{F}_q$, as desired. Let $z \neq \gamma$. Since $z = X^n z X^{-n} = \sigma^n(z)$ for all $n \in \mathbb{Z}$, and since f and g are coprime, we see that f and g are equal to finite products of the form $\prod \alpha_{\mathfrak{D}}^{n(\mathfrak{D})}$ with $n(\mathfrak{D}) \geq 0$. Then also $\gamma \in Z(B)$, whence $\gamma \in \mathbb{F}_q$.

(iii) We have $Z(\mathfrak{A}_1) \subset Z(B)$, and the required result follows from the preceding one. ∎

The main properties of ideals of $\mathfrak{A}_1$ are summarized in the following theorem.

Theorem 5.16 (i) *Every nonzero prime ideal of $\mathfrak{A}_1$ is a maximal ideal.*

(ii) *Each nonzero ideal of $\mathfrak{A}_1$ is a unique finite product of maximal ideals. All ideals commute.*

(iii) *The mapping* $Cyc \to \text{Specm}(\mathfrak{A}_1)$,

$$\mathfrak{O} \mapsto \mathfrak{m}_\mathfrak{O} \stackrel{def}{=} \mathfrak{A}_1 \prod_{\mathfrak{p} \in \mathfrak{O}} \mathfrak{p} = \mathfrak{A}_1 \alpha_\mathfrak{O},$$

is a bijection with inverse $\mathfrak{m} \mapsto \text{Supp}(\mathfrak{A}_1/\mathfrak{m})$.

For the proof and further details regarding the ideal structure and other algebraic properties of $\mathfrak{A}_1$ see [11].

5.4 Quasi-holonomic modules

5.4.1. Dimensions. Let M be a filtered left module with a good filtration $\{M_\nu\}$ over the Carlitz ring $\mathfrak{A}_{n+1}$ (the filtration in $\mathfrak{A}_{n+1}$ was introduced in Section 5.2.2). Below we always assume that the submodules M_ν are finite-dimensional as vector spaces over $\overline{K}_c$.

Let us consider the associated graded module

$$\text{gr}\, M = \bigoplus_{\nu=0}^{\infty} (\text{gr}\, M)_\nu, \quad (\text{gr}\, M)_\nu = M_\nu/M_{\nu-1} \ (M_{-1} = \{0\}).$$

As we know, the good filtration assumption means that $\text{gr}\, M$ is finitely generated.

The next results show that some of the basic properties of filtered and graded modules over $\mathfrak{A}_{n+1}$ resemble those for modules over the Weyl algebras [16, 30], though a contrasting phenomenon found in Theorem 5.14 will lead to some different properties too.

It will be convenient to consider, together with the rings $\mathfrak{A}_{n+1}$, $n \geq 0$, the ring $\mathfrak{A}_0$ generated by τ and the scalars $\lambda \in \overline{K}_c$, and the modules over $\mathfrak{A}_0$.

In the proof of the next theorem we will use an auxiliary result (see [30]). A polynomial $p \in \mathbb{Q}[t]$ is called *numerical* if it takes values from $\mathbb{Z}$ at all sufficiently big natural numbers.

Lemma 5.17 (i) *Let* $f : \mathbb{Z} \to \mathbb{Z}$ *be a function such that* $f(l) - f(l-1) = \gamma(l)$ *for* $l \geq l_0 > 0$, *where* $\gamma(l)$ *is a numerical polynomial. Then there exists a number* $l_1 > 0$ *such that* f *coincides, for* $l \geq l_1$, *with some numerical polynomial.*

(ii) *Let* $\gamma(t)$ *be a numerical polynomial. Then there exist integers*

$c_0, \ldots, c_k$ such that
$$\gamma(t) = \sum_{i=0}^{k} c_{k-i} \binom{t}{i}$$
where $\binom{t}{i} = \dfrac{t(t-1)\cdots(t-i+1)}{i!}$. In particular, $\gamma(\nu) \in \mathbb{Z}$ for all $\nu \in \mathbb{Z}$.

Theorem 5.18 *There exist a polynomial $\chi \in \mathbb{Q}[t]$, and a positive number σ such that*
$$\sum_{i=0}^{s} \dim(M_i/M_{i-1}) = \chi(s) \quad \text{for } s \geq \sigma.$$
Here dim *means the dimension over* $\overline{K}_c$.

Proof. First we consider the case $n = -1$, that is filtered and graded $\mathfrak{A}_0$-modules. Let $F_0 : \mathbb{Z} \to \mathbb{Z}$,
$$F_0(s) = \sum_{i=-\infty}^{s} \dim \mathfrak{M}_i$$
where $\mathfrak{M}_i = M_i/M_{i-1}$, as $i \geq 0$, $\mathfrak{M}_i = \{0\}$, as $i < 0$.

The graded ring $\mathrm{gr}(\mathfrak{A}_0)$ is generated (compare Section 5.2.2) by scalars $\lambda \in \overline{K}_c$ and the element $\overline{\tau}$. Let $\varphi_{i-1} : \mathfrak{M}_{i-1} \to \mathfrak{M}_i$, $\varphi_i(m) = \overline{\tau}(m)$, $m \in \mathfrak{M}_{i-1}$. Denote $Q_{i-1} = \ker \varphi_i$, $L_i = \mathfrak{M}_i/\operatorname{Im} \varphi_{i-1}$ and consider the exact sequence
$$0 \longrightarrow Q_{i-1} \longrightarrow \mathfrak{M}_{i-1} \overset{\varphi_{i-1}}{\longrightarrow} \mathfrak{M}_i \longrightarrow L_i \longrightarrow 0 \qquad (5.16)$$
of vector spaces over $\overline{K}_c$.

The mappings φ_i are semilinear with respect to τ considered as an automorphism of the algebraically closed field $\overline{K}_c$ (see [18]). As explained in Appendix 1 to Chapter 2 of [18], basic notions of linear algebra remain valid for semilinear mappings – a semilinear mapping of a vector space into itself can be interpreted as a linear mapping between two different vector spaces, and, for instance, dimensions of the kernel and cokernel are not changed in this interpretation. Now we find from (5.16) that
$$\dim \mathfrak{M}_{i-1} = \dim Q_{i-1} + \dim (\operatorname{Im} \varphi_{i-1}),$$
$$\dim \mathfrak{M}_i = \dim (\operatorname{Im} \varphi_{i-1}) + \dim L_i,$$

whence
$$\dim Q_{i-1} - \dim \mathfrak{M}_{i-1} + \dim \mathfrak{M}_i - \dim L_i = 0. \tag{5.17}$$

Note that $Q = \bigoplus_{i \geq 0} Q_i$ and $L = \bigoplus_{i \geq 0} L_i$ are graded finitely generated modules over $\mathrm{gr}(\mathfrak{A}_0)$ which are annihilated by the operator $\overline{\tau}$. Therefore Q and L are finite-dimensional vector spaces over $\overline{K}_c$, thus $Q_i = L_i = \{0\}$ for i big enough, so that $\sum_{i=0}^{s} \dim Q_i = \mathrm{const}$, $\sum_{i=0}^{s} \dim L_i = \mathrm{const}$, if s is big enough. Summing up the identities (5.17) we find $\sigma_0 > 0$ such that

$$F_0(s) - F_0(s-1) = \mathrm{const} \in \mathbb{Z}, \quad s \geq \sigma_0,$$

and $F_0(s)$ coincides, for s big enough, with a numerical polynomial of degree 1.

Let $n = 0$, that is we will consider $\mathfrak{A}_1$-modules. We repeat the above reasoning with the mappings $\psi_i : \mathfrak{M}_{i-1} \to \mathfrak{M}_i$, $\psi_i(m) = \overline{d}(m)$, semilinear with respect to the automorphism τ^{-1}. If $F_1(s) = \sum_{i=-\infty}^{s} \dim \mathfrak{M}_i$ (the notation F_1 corresponds to $\mathfrak{A}_1$-modules), we find, as above, on the basis of the result for $\mathfrak{A}_0$, that

$$F_1(s) - F_1(s-1) = \gamma(s), \quad s \geq \sigma_1 > 0,$$

where $\gamma(s)$ is a numerical polynomial, and it remains to use Lemma 5.17.

The same reasoning (for the linear mappings corresponding to $\overline{\Delta}_1, \ldots, \overline{\Delta}_n$), with induction on n, brings the required result for the general situation. ∎

The number $d(M) = \deg \chi$ is called the *dimension* of M, while the leading coefficient of χ multiplied by $d(M)!$ is called the *multiplicity* of M. It follows from Lemma 5.17 (ii) that the multiplicity (just as the dimension) is always a natural number. These definitions follow the notions related to modules over the Weyl algebras and their generalizations [16, 30, 73]; just as for modules over algebras (see Lemma 6.1 in [73]), it can be shown that $d(M)$ agrees with the general notion of the Gelfand–Kirillov dimension. Following the existing tradition, we call χ the *Hilbert polynomial*. Note that $\chi(i) = \dim M_i$ for big enough values of i.

Below we return to the earlier notation considering the rings $\mathfrak{A}_{n+1}$ for $n \geq 0$.

Let us look more closely at the notion of a good filtration. Recall that the filtration $\{\Gamma_\nu\}$ in $\mathfrak{A}_{n+1}$ was introduced in Section 5.2.2.

Proposition 5.19 *Let M be a filtered module with a filtration $\{M_\nu\}$ over $\mathfrak{A}_{n+1}$. The filtration $\{M_\nu\}$ is good if and only if there exists a number k_0 such that*

$$M_{i+k} = \Gamma_i M_k, \text{ for all } i \in \mathbb{Z}_+, k \geq k_0. \tag{5.18}$$

Proof. Suppose that the condition (5.18) holds. Since $\dim M_{k_0} < \infty$, M_{k_0} possesses a finite $\overline{K}_c$-basis, and images of its elements generate $\mathrm{gr}(M)$. Thus the filtration is good.

Conversely, let $\mathrm{gr}(M)$ be finitely generated, with generators represented by the elements $u_1, \ldots, u_s$, where $u_j \in M_{k_j} \setminus M_{k_j-1}$, $j = 1, 2, \ldots, s$. Set

$$k_0 = \max\{k_1, \ldots, k_s\}.$$

We will prove (5.18) by induction on i. The case $i = 0$ is trivial. Suppose that

$$M_{i-1+k} = \Gamma_{i-1} M_k, \quad k \geq k_0, \tag{5.19}$$

for some i. Let $v \in M_{i+k}$. Denote by μ_l and σ_l the canonical projections

$$\mu_l : M_l \to M_l/M_{l-1}, \quad \sigma_l : \Gamma_l \to \Gamma_l/\Gamma_{l-1}.$$

By the finite generation assumption,

$$\mu_{i+k}(v) \in \sum_{j=1}^s \sigma_{i+k-k_j}\left(\Gamma_{i+k-k_j}\right) \mu_{k_j}(u_j).$$

Since $\Gamma_{i+k-k_j} = \Gamma_i \cdot \Gamma_{k-k_j}$ (that is every element from the left-hand set is a sum of products of elements from Γ_i and Γ_{k-k_j}), we see that

$$v \in \sum_{j=1}^s \Gamma_i \cdot \Gamma_{k-k_j} u_j + M_{i+k-1}.$$

Now we have $\Gamma_{k-k_j} u_j \in M_k$, and it remains to use (5.19) to obtain (5.18). ∎

The next results deal with the comparison of filtrations.

Proposition 5.20 *Suppose that M is a left $\mathfrak{A}_{n+1}$-module with two filtration, $\left\{M_i^{(1)}\right\}$ and $\left\{M_i^{(2)}\right\}$.*

(i) If $\left\{M_i^{(1)}\right\}$ is a good filtration, then there exists a number k_1 such that

$$M_i^{(1)} \subset M_{i+k_1}^{(2)}, \text{ for all } i. \tag{5.20}$$

(ii) *If both filtrations are good, then there exists a number k_2 such that*

$$M^{(2)}_{i-k_2} \subset M^{(1)}_i \subset M^{(2)}_{i+k_2}, \text{ for } i \text{ big enough.} \tag{5.21}$$

Proof. If $\left\{M^{(1)}_i\right\}$ is good, then there exists a number k_0 such that $\Gamma_i M^{(1)}_j = M^{(1)}_{i+j}$, for all $j \geq k_0$, $i \geq 0$. Since $M^{(1)}_{k_0}$ is finite-dimensional over $\overline{K}_c$, we find that $M^{(1)}_{k_0} \subset M^{(2)}_{k_1}$ for some k_1. Then

$$M^{(1)}_{i+k_0} = \Gamma_i \cdot M^{(1)}_{k_0} \subset \Gamma_i \cdot M^{(2)}_{k_1} \subset M^{(2)}_{i+k_1},$$

so that

$$M^{(1)}_i \subset M^{(1)}_{i+k_0} \subset M^{(2)}_{i+k_1}, \quad i \geq 0,$$

and we have proved (5.20).

The second assertion is a consequence of the first one. ∎

Corollary 5.21 *The dimension $d(M)$ and the multiplicity $m(M)$ are the same for any good filtration of a $\mathfrak{A}_{n+1}$-module M.*

Proof. Suppose we have two good filtrations, as in Proposition 5.20 (ii). Let $\chi^{(1)}$ and $\chi^{(2)}$ be the corresponding Hilbert polynomials. It follows from (5.21) that

$$\chi^{(2)}(i - k_2) \leq \chi^{(1)}(i) \leq \chi^{(2)}(i + k_2)$$

for large values of i. Since the behavior of a polynomial at infinity is determined by its degree, we get that the degrees and leading coefficients of the polynomials $\chi^{(1)}$ and $\chi^{(2)}$ coincide. ∎

Let us consider the behavior of the dimension $d(M)$ in the context of some operations on $\mathfrak{A}_{n+1}$-modules.

Proposition 5.22 (i) *Let M be a filtered $\mathfrak{A}_{n+1}$-module with a good filtration, $N \subset M$ a submodule, M/N a quotient module (with the induced filtrations). Then*

$$d(M) = \max\{d(N), d(M/N)\}, \tag{5.22}$$

and if $d(N) = d(M/N)$, then

$$m(M) = m(N) + m(M/N). \tag{5.23}$$

(ii) Let $L_1, \ldots, L_k$ be filtered $\mathfrak{A}_{n+1}$-modules with good filtrations, $L = \bigoplus_{j=1}^{k} L_j$. Then

$$d(L) = \max\{d(L_1), \ldots, d(L_k)\}, \tag{5.24}$$

and if $d(L) = d(L_i)$ for all i, then

$$m(L) = \sum_{i=1}^{k} m(L_i). \tag{5.25}$$

Proof. (i) Let $\{N_k\}$ and $\{F_k\}$ be the induced filtrations in N and M/N respectively (see Section 5.1.2). By (5.1), for each k we have the exact sequence

$$0 \to N_k/N_{k-1} \to M_k/M_{k-1} \to F_k/F_{k-1} \to 0$$

which implies the equality

$$\dim N_k/N_{k-1} + \dim F_k/F_{k-1} = \dim M_k/M_{k-1}.$$

Summing up by k, we obtain the equality for the corresponding Hilbert polynomials

$$\chi_N(s) + \chi_{M/N}(s) = \chi_M(s)$$

valid for large s, thus for all s. This implies (5.22) and (5.23).

(ii) The proof of (5.24) and (5.25) follows by induction from part (i) applied to the exact sequence

$$0 \to L_k \to L \to L_1 \oplus \cdots \oplus L_{k-1} \to 0. \quad \blacksquare$$

5.4.2. Dimension: calculations and estimates. First of all, consider $\mathfrak{A}_{n+1}$ as a left module over itself. It follows from Lemma 5.10 that

$$d(\mathfrak{A}_{n+1}) = n+2, \quad m(\mathfrak{A}_{n+1}) = 1. \tag{5.26}$$

Next, let us consider direct sums

$$\mathfrak{A}_{n+1}^r = \underbrace{\mathfrak{A}_{n+1} \oplus \cdots \oplus \mathfrak{A}_{n+1}}_{r}.$$

By (5.24), we find that $d\left(\mathfrak{A}_{n+1}^r\right) = n+2$ for any $r = 1, 2, \ldots$. This leads to the following general result.

Proposition 5.23 *For any finitely generated left $\mathfrak{A}_{n+1}$-module M,*

$$d(M) \leq n + 2. \tag{5.27}$$

Proof. Let M be generated by r elements. Then there exists an epimorphism $\varphi : \mathfrak{A}_{n+1}^r \to M$, thus $M \cong \mathfrak{A}_{n+1}^r / \ker \varphi$, and, by virtue of Proposition 5.22 (i),

$$d\left(\mathfrak{A}_{n+1}^r\right) = \max\{d(M), d(\ker \varphi)\},$$

whence $d(M) \leq d\left(\mathfrak{A}_{n+1}^r\right) = n + 2$. ∎

By (5.26), in general the bound (5.27) cannot be improved. Nevertheless, there are cases with smaller values of $d(M)$. An important special case is given by the next proposition.

Proposition 5.24 *If I is a nonzero left ideal in $\mathfrak{A}_{n+1}$, then*

$$d\left(\mathfrak{A}_{n+1}/I\right) \leq n + 1. \tag{5.28}$$

Proof. First we consider the case of a principal left ideal $I = \mathfrak{A}_{n+1}\gamma$, $\gamma \in \mathfrak{A}_{n+1}$. Then we have an exact sequence

$$0 \to \mathfrak{A}_{n+1} \xrightarrow{\theta} \mathfrak{A}_{n+1} \to \mathfrak{A}_{n+1}/\mathfrak{A}_{n+1}\gamma \to 0$$

where $\theta(a) = a\gamma$ for all $a \in \mathfrak{A}_{n+1}$. Here we used the fact (Theorem 5.9) that $\mathfrak{A}_{n+1}$ has no zero-divisors.

Now, if $d\left(\mathfrak{A}_{n+1}/I\right) = n + 2 = d\left(\mathfrak{A}_{n+1}\right)$, then, by (5.23), we would have

$$m\left(\mathfrak{A}_{n+1}\right) = m(I) + m\left(\mathfrak{A}_{n+1}/I\right)$$

which contradicts the equality $m\left(\mathfrak{A}_{n+1}\right) = 1$, since the multiplicity is a natural number. ∎

Let us consider the set $\widehat{\mathcal{F}}_{n+1}$ of polynomials (5.2) as a $\mathfrak{A}_{n+1}$-module. A filtration

$$\mathcal{F}_{n+1}^{(0)} \subset \mathcal{F}_{n+1}^{(1)} \subset \ldots \subset \widehat{\mathcal{F}}_{n+1}$$

can be introduced by setting $\mathcal{F}_{n+1}^{(j)}$ to be the collection of all the polynomials (5.2), in which the maximal indices $k_1, \ldots, k_n$ corresponding to nonzero coefficients $a_{m, k_1, \ldots, k_n}$ do not exceed j. This filtration is obviously good.

Proposition 5.25 *For the module* $\widehat{\mathcal{F}}_{n+1}$,

$$d\left(\widehat{\mathcal{F}}_{n+1}\right) = n+1, \quad m\left(\widehat{\mathcal{F}}_{n+1}\right) = n! \tag{5.29}$$

Proof. Let us compute $\dim \mathcal{F}_{n+1}^{(j)}$. For a fixed μ, the quantity of n-tuples $(k_1, \ldots, k_n)$ of nonnegative integers, for which $\min(k_1, \ldots, k_n) = \mu$, is added up from those n-tuples where i numbers are equal to μ while $n-i$ numbers are strictly larger and can take $j - \mu$ values. Therefore the above quantity equals $\sum_{i=1}^{n} \binom{n}{i}(j-\mu)^{n-i}$. Next, $\mu + 1$ possible values of m in (5.2) correspond to each n-tuple. Thus,

$$\dim \mathcal{F}_{n+1}^{(j)} = \sum_{\mu=0}^{j}(\mu+1) \sum_{i=1}^{n} \binom{n}{i}(j-\mu)^{n-i}$$

$$= \sum_{\mu=0}^{j}(\mu+1) \left\{(j-\mu+1)^n - (j-\mu)^n\right\}.$$

Denote $r_\mu = (j-\mu+1)^n - (j-\mu)^n$, $R_i = r_0 + r_1 + \cdots + r_i = (j+1)^n - (j-i)^n$. Performing the Abel transformation we get

$$\dim \mathcal{F}_{n+1}^{(j)} = (j+1)R_j - \sum_{i=0}^{j-1} R_i$$

$$= (j+1)^{n+1} - j(j+1)^n + \sum_{i=0}^{j-1}(j-i)^n$$

$$= (j+1)^n + \sum_{k=1}^{j} k^n = (j+1)^n + S_n(j+1)$$

where $S_n(N) = 1^n + 2^n + \cdots + (N-1)^n$.

It is known ([52], Chapter 15) that

$$S_n(N) = \frac{1}{n+1} \sum_{k=0}^{n} \binom{n+1}{k} B_k N^{n+1-k}$$

where B_k are the Bernoulli numbers. Therefore we find that

$$\dim \mathcal{F}_{n+1}^{(j)} = \frac{(j+1)^{n+1}}{n+1} + P_n(j)$$

where P_n is a polynomial of degree n. This implies (5.29). ∎

It is instructive to compare the above calculations with those for modules over the Weyl algebras [16, 25, 30]. For such modules, possible values of $d(M)$ are also bounded from above by the dimension of the algebra considered as a module over itself. Moreover, by the celebrated Bernstein inequality, $d(M)$ is also bounded from below by the dimension of the module consisting of polynomials. Modules with the minimal possible dimension are called holonomic, and it appears that just such modules emerge in many important applications.

In our situation, an analog of the Bernstein inequality is impossible, since there are examples of finite-dimensional $\mathfrak{A}_1$-modules M (see Theorem 5.14), for which obviously $d(M) = 0$. However, the above analogy makes it natural to call an $\mathfrak{A}_{n+1}$-module M *quasiholonomic*, if $d(M) = n+1$. As we will see, lower dimensions correspond in applications to degenerate cases, while quasiholonomic modules play roles similar to those of holonomic modules over the Weyl algebras.

Returning to possible values of $d(M)$ for $\mathfrak{A}_1$-modules, we mention a special case opposite to the one considered in Theorem 5.14.

Proposition 5.26 *Let M be a finitely generated $\mathfrak{A}_1$-module with a good filtration. Suppose that there exists a "vacuum vector" $v \in M$, such that $d_s v = 0$ and $\tau^m(v) \neq 0$ for all $m = 0, 1, 2, \ldots$. Then $d(M) \geq 1$.*

Proof. It follows from (5.12) that
$$d_s \tau^m v = [m]^{1/q} \tau^{m-1} v, \quad m = 1, 2, \ldots,$$
that is $\tau^{m-1} v$ is an eigenvector of a linear operator $d_s \tau$ on M (considered as a $\overline{K}_c$-vector space) corresponding to the eigenvalue $[m]^{1/q}$. Therefore the vectors $\tau^{m-1} v$ are linearly independent. It follows from the existence of the Hilbert polynomial χ implementing the dimension $d(M)$ that $d(M) \geq 1$. ∎

5.4.3. Evolution operators. Let $R \in \mathfrak{A}_{n+1}$ be an operator of the form
$$R = P(\Delta_1, \ldots, \Delta_n) + Q(\Delta_1, \ldots, \Delta_n) d$$
where P, Q are non-zero polynomials. As we know (Theorem 3.10), under some nondegeneracy conditions the Cauchy problem for the equation $Ru = 0$ is well-posed. Let $I = \mathfrak{A}_{n+1} R$.

Theorem 5.27 *The module $M = \mathfrak{A}_{n+1}/I$ is quasiholonomic.*

Proof. Due to (5.28), we have to show only that $d(M) \geq n+1$. First we prove two lemmas.

Lemma 5.28 *An operator*
$$A = \sum a_{l,\mu,i_1,\ldots,i_n} \tau^l d^\mu \Delta_1^{i_1} \ldots \Delta_n^{i_n}, \quad a_{l,\mu,i_1,\ldots,i_n} \in \overline{K}_c, \tag{5.30}$$
is linear if and only if $a_{l,\mu,i_1,\ldots,i_n} = 0$ for $l \neq \mu$.

Proof. Let $\sigma \in \overline{K}_c$. Suppose that $A\sigma = \sigma A$, that is
$$\sum \sigma a_{l,\mu,i_1,\ldots,i_n} \tau^l d^\mu \Delta_1^{i_1} \ldots \Delta_n^{i_n} = \sum a_{l,\mu,i_1,\ldots,i_n} \sigma^{q^{l-\mu}} \tau^l d^\mu \Delta_1^{i_1} \ldots \Delta_n^{i_n}.$$
By the uniqueness of the representation (5.30) (Proposition 5.8), we find that $\sigma^{q^{l-\mu}} = \sigma$, whenever $a_{l,\mu,i_1,\ldots,i_n} \neq 0$. Since σ is arbitrary, that is possible if and only if $l = \mu$. ∎

Lemma 5.29 *The ideal I does not contain nonzero linear operators.*

Proof. Using the commutation relations defining $\mathfrak{A}_{n+1}$ we can rewrite the representation (5.30) of an arbitrary operator $A \in \mathfrak{A}_{n+1}$ in the form
$$A = \sum a'_{l,\mu,i_1,\ldots,i_n} \Delta_1^{i_1} \ldots \Delta_n^{i_n} \tau^l d^\mu, \quad a'_{l,\mu,i_1,\ldots,i_n} \in \overline{K}_c. \tag{5.31}$$
Just as in (5.30), the coefficients $a'_{l,\mu,i_1,\ldots,i_n}$ are determined in a unique way, and Lemma 5.28 remains valid for the representation (5.31).

Suppose that an operator $A \in \mathfrak{A}_{n+1}$ is such that AR is linear. Let us write (5.31) in the form
$$A = \sum_{l=0}^{N} \sum_{\mu=0}^{N} \alpha_{l\mu} \tau^l d^\mu$$
where $\alpha_{l\mu}$ are the appropriate elements of the commutative $\overline{K}_c$-algebra $\mathcal{D}$ (without zero-divisors) generated by the linear operators $\Delta_1, \ldots, \Delta_n$. We have also $R = \gamma + \delta d$, $\gamma, \delta \in \mathcal{D}$, $\gamma \neq 0$, $\delta \neq 0$. In this new notation,
$$AR = \left(\sum_{l=0}^{N} \sum_{\mu=0}^{N} \alpha_{l\mu} \tau^l d^\mu \right) (\gamma + \delta d).$$

As an element $\gamma \in \mathcal{D}$ is permuted with powers of τ and d, in additional terms (appearing in accordance with the commutation rules) the powers of

τ and d respectively are the same, while the degrees of elements from $\mathcal{D}$ (as polynomials of $\Delta_1, \ldots, \Delta_n$) decrease. Therefore

$$AR = \sum_{l=0}^{N} \sum_{\mu=0}^{N} \alpha_{l\mu} \gamma' \tau^l d^\mu + \sum_{l=0}^{N} \sum_{\mu=0}^{N} \alpha_{l\mu} \delta' \tau^l d^{\mu+1}$$

where $\gamma', \delta' \in \mathcal{D}$, $\gamma' \neq 0, \delta' \neq 0$, whence

$$AR = \sum_{l=0}^{N} \sum_{\nu=1}^{N} (\alpha_{l\nu}\gamma' + \alpha_{l,\nu-1}\delta') \tau^l d^\nu + \sum_{l=0}^{N} \left(\alpha_{l0}\gamma' \tau^l + \alpha_{lN}\delta' \tau^l d^{N+1} \right).$$

By Lemma 5.28,

$$\alpha_{l0} = 0, \quad l = 1, \ldots, N; \tag{5.32}$$
$$\alpha_{lN} = 0, \quad l = 0, 1, \ldots, N. \tag{5.33}$$

Considering terms with $l < N, \nu = N$, we find that

$$\alpha_{lN}\gamma' + \alpha_{l,N-1}\delta' = 0,$$

and (5.33) yields $\alpha_{l,N-1} = 0$, $0 \le l \le N-1$. Repeating the reasoning we obtain that $\alpha_{l\nu} = 0$ for $l \le \nu$.

On the other hand, for $l \ge 2, \nu = 1$ we get

$$\alpha_{l1}\gamma' + \alpha_{l0}\delta' = 0,$$

and, by (5.32), $\alpha_{l1} = 0$. Repeating we come to the conclusion that $\alpha_{l\nu} = 0$ for $l > \nu$, so that $A = 0$. ∎

Proof of Theorem 5.27 (continued). Let us consider the induced filtration in M. The subspace M_ν is generated by images in M of the elements (5.30) with $\max(l + \mu + i_1 + \cdots + i_n) \le \nu$; those two elements whose difference belongs to I are identified. Let us consider elements with $l = \mu$.

Elements of the form $\tau^l d^l \Delta_1^{i_1} \cdots \Delta_n^{i_n}$ with different collections of parameters $(l, i_1, \ldots, i_n)$ are linearly independent in $\mathfrak{A}_{n+1}$. If some linear combination of their images equals zero in M, then the corresponding linear combination of the elements themselves must belong to I, which (by Lemma 5.29) is possible only if it is equal to zero. Therefore the images of the above elements are linearly independent, so that

$$\dim M_\nu \ge \operatorname{card} \left\{ (l, i_1, \ldots, i_n) \in \mathbb{Z}_+^{n+1} : 2l + i_1 + \cdots + i_n \le \nu \right\}$$
$$\ge \operatorname{card} \left\{ (l, i_1, \ldots, i_n) \in \mathbb{Z}_+^{n+1} : l + i_1 + \cdots + i_n \le \operatorname{int}(\nu/2) \right\}.$$

Evaluating the number of nonnegative integral solutions of the above inequality as in the proof of Lemma 5.10 we find that

$$\dim M_\nu \geq \binom{[\nu/2]+n+1}{n+1} \geq c_1 [\nu/2]^{n+1} \geq c_2 \nu^{n+1}$$

for large values of ν (c_1, c_2 are positive constants independent of ν). Thus, $d(M) \geq n+1$, as desired. ∎

5.4.4. Quasiholonomic functions. Let $0 \neq f \in \mathcal{F}_{n+1}$,

$$I_f = \{\varphi \in \mathfrak{A}_{n+1} : \varphi(f) = 0\}.$$

I_f is a left ideal in $\mathfrak{A}_{n+1}$. The left $\mathfrak{A}_{n+1}$-module $M_f = \mathfrak{A}_{n+1}/I_f$ is isomorphic to the submodule $\mathfrak{A}_{n+1} f \subset \mathcal{F}_{n+1}$ – an element $\varphi(f) \in \mathfrak{A}_{n+1} f$ corresponds to the class of $\varphi \in \mathfrak{A}_{n+1}$ in M_f. A natural good filtration in M_f is induced from that in $\mathfrak{A}_{n+1}$ – the subspace M_j is generated by elements $\tau^l d_s^\mu \Delta_1^{i_1} \ldots \Delta_n^{i_n} f$ with $l + \mu + i_1 + \cdots + i_n \leq j$.

As we know, if $I_f \neq \{0\}$, then $d(M_f) \leq n+1$. We call a function f *quasiholonomic* if the module M_f is quasiholonomic, that is $d(M_f) = n+1$. The condition $I_f \neq \{0\}$ means that f is a solution of a "differential equation" $\varphi(f) = 0$, $\varphi \in \mathfrak{A}_{n+1}$. For $n = 0$, we have the following easy result.

Theorem 5.30 *If a nonzero function $f \in \mathcal{F}_1$ satisfies an equation $\varphi(f) = 0$, $0 \neq \varphi \in \mathfrak{A}_1$, then f is quasiholonomic.*

Proof. It is sufficient to show that $\dim M_f = \infty$. In fact, the sequence $\{\tau^l f\}_{l=0}^\infty$ is linearly independent because otherwise we would have a finite collection of elements $c_0, c_1, \ldots, c_N \in \overline{K}_c$, some of which are different from zero, such that

$$c_0 f(s) + c_1 f^q(s) + \cdots + c_N f^{q^N}(s) = 0 \tag{5.34}$$

for all s from a neighborhood of the origin in $\overline{K}_c$. It follows from (5.34) that f takes only a finite number of values. By the uniqueness theorem for non-Archimedean holomorphic functions, $f(s) \equiv \text{const}$ on some neighborhood of the origin. Due to the $\mathbb{F}_q$-linearity, $f(s) \equiv 0$, and we have come to a contradiction. ∎

In particular, any $\mathbb{F}_q$-linear polynomial of s is quasiholonomic, since it is annihilated by d_s^m, with a sufficiently large m.

If $n > 0$, the situation is more complicated. We call the module M_f

(and the corresponding function f) *degenerate* if $d(M_f) < n+1$ (by the Bernstein inequality, there are no degeneracy phenomena for modules over the complex Weyl algebra). We give an example of degeneracy for the case $n = 1$.

Let $f(s, t_1) = g(st_1) \in \mathcal{F}_2$ where the function g belongs to $\mathcal{F}_1$ and satisfies an equation $\varphi(g) = 0$, $\varphi \in \mathfrak{A}_1$. Then f is degenerate.

Indeed, by the general rule, $\mathfrak{M}_j$ is spanned by elements $\tau^l d_s^\mu \Delta_1^{i_1} f$ with $l + \mu + i_1 \leq j$. In the present situation,

$$\Delta_1 f = g(xst_1) - xg(st_1) = \tau d_s g,$$

so that an element $\tau^l d_s^\mu \Delta_1^{i_1} f$ is a linear combination of elements $\left(\tau^{l+\lambda} d_s^{\mu+\nu} g\right)(s,t)$ with $\lambda \leq i_1$, $\nu \leq i_1$. Therefore $\mathfrak{M}_j$ is contained in the linear hull of elements $\tau^k d_s^m g$, $k + m \leq 2j$. By Theorem 5.30, the $\overline{K}_c$-dimension of the latter does not exceed a linear function of $2j$, so that $d(M_f) \leq 1$. On the other hand, since, as in the proof of Theorem 5.30, the system of functions $\{\tau^l f\}_{l=0}^\infty$ is linearly independent, we find that $d(M_f) = 1$.

In order to exclude the degenerate case, we introduce the notion of a nonsparse function.

A function $f \in \mathcal{F}_{n+1}$ of the form (5.2) is called *nonsparse* if there exists a sequence $m_l \to \infty$ such that, for any l, there exist sequences $k_1^{(i)}, k_2^{(i)}, \ldots, k_n^{(i)} \geq m_l$ (depending on l), such that $k_\nu^{(i)} \to \infty$ as $i \to \infty$ ($\nu = 1, \ldots, n$), and $a_{m_l, k_1^{(i)}, \ldots, k_n^{(i)}} \neq 0$.

Lemma 5.31 *If a function f is nonsparse, then the system of functions $(\tau d_s)^\lambda \Delta_1^{j_1} \ldots \Delta_n^{j_n} f$ ($\lambda, j_1, \ldots, j_n = 0, 1, 2, \ldots$) is linearly independent over $\overline{K}_c$.*

Proof. Suppose that

$$\sum_{\lambda=0}^\Lambda \sum_{j_1=0}^{J_1} \cdots \sum_{j_n=0}^{J_n} c_{\lambda, j_1, \ldots, j_n} (\tau d_s)^\lambda \Delta_1^{j_1} \ldots \Delta_n^{j_n} f = 0 \qquad (5.35)$$

for some $c_{\lambda, j_1, \ldots, j_n} \in \overline{K}_c$, $\Lambda, J_1, \ldots, J_n \in \mathbb{N}$. Substituting (5.2) into (5.35) and collecting coefficients of the power series we find that

$$\sum_{\lambda=0}^\Lambda \sum_{j_1=0}^{J_1} \cdots \sum_{j_n=0}^{J_n} c_{\lambda, j_1, \ldots, j_n} [m_l]^\lambda [k_1^{(i)}]^{j_1} \ldots [k_n^{(i)}]^{j_n} = 0 \qquad (5.36)$$

for all l, i.

We see from (5.36) that the polynomial

$$\sum_{j_n=0}^{J_n}\left\{\sum_{\lambda=0}^{\Lambda}\sum_{j_1=0}^{J_1}\cdots\sum_{j_{n-1}=0}^{J_{n-1}} c_{\lambda,j_1,\ldots,j_n}[m_l]^\lambda[k_1^{(i)}]^{j_1}\cdots[k_{n-1}^{(i)}]^{j_{n-1}}\right\}z^{j_n}$$

has an infinite sequence of different roots, so that

$$\sum_{\lambda=0}^{\Lambda}\sum_{j_1=0}^{J_1}\cdots\sum_{j_{n-1}=0}^{J_{n-1}} c_{\lambda,j_1,\ldots,j_n}[m_l]^\lambda[k_1^{(i)}]^{j_1}\cdots[k_{n-1}^{(i)}]^{j_{n-1}} = 0$$

for all l, i, and for each $j_n = 0, 1, \ldots, J_n$. Repeating this reasoning we find that all the coefficients $c_{\lambda,j_1,\ldots,j_n}$ are equal to zero. ∎

Now the above arguments regarding $d(M_f)$ yield the following result.

Theorem 5.32 *If a function f is nonsparse, then $d(M_f) \geq n+1$. If, in addition, f satisfies an equation $\varphi(f) = 0$, $0 \neq \varphi \in \mathfrak{A}_{n+1}$, then f is quasiholonomic.*

As in the classical situation, one can construct quasiholonomic functions by addition.

Proposition 5.33 *If the functions $f, g \in \mathcal{F}_{n+1}$ are quasiholonomic, and $f + g$ is nonsparse, then $f + g$ is quasiholonomic.*

Proof. Consider the $\mathfrak{A}_{n+1}$-module $M_2 = (\mathfrak{A}_{n+1}f) \oplus (\mathfrak{A}_{n+1}g)$. Since f and g are both quasiholonomic, we have $d(M_2) = n+1$. Next, let N_2 be a submodule of M_2 consisting of such pairs $(\varphi(f), \varphi(g))$ that $\varphi(f)+\varphi(g) = 0$. Then $d(M_2) = \max\{d(N_2), d(M_2/N_2)\}$, so that $d(M_2/N_2) \leq n+1$.

On the other hand, we have an injective mapping $\mathfrak{A}_{n+1}(f+g) \to M_2/N_2$, which maps $\varphi(f+g)$ to the image of $(\varphi(f), \varphi(g))$ in M_2/N_2. Therefore $d(\mathfrak{A}_{n+1}(f+g)) \leq d(M_2/N_2) \leq n+1$. It remains to use Theorem 5.32. ∎

It was noticed by Zeilberger (see [25] and references therein) that many classical sequences of functions generate holonomic modules, if, as a preparation, a transform is made with respect to the discrete variables, reducing the continuous–discrete case to the purely continuous one (simultaneously in all the variables). In our situation, if, for example, $\{P_k(s)\}$ is a sequence of $\mathbb{F}_q$-linear polynomials with $\deg P_k \leq q^k$, we set

$$f(s,t) = \sum_{k=0}^{\infty} P_k(s) t^{q^k}. \tag{5.37}$$

Then $f \in \mathcal{F}_2$, and we may consider the question of whether f is quasiholonomic. It is remarkable that several functions obtained this way from the basic special functions of $\mathbb{F}_q$-linear analysis, are indeed quasiholonomic.

(a) *The Carlitz polynomials.* Considering the construction (5.37) for $P_k = f_k$, the normalized Carlitz polynomials, we obtain the Carlitz module (Section 1.6.2), that is

$$C_s(t) = \sum_{k=0}^{\infty} f_k(s) t^{q^k}. \tag{5.38}$$

By Proposition 1.7,

$$f_k(s) = \sum_{i=0}^{k} \frac{(-1)^{k-i}}{D_i L_{k-i}^{q^i}} s^{q^i}.$$

It is easy to check that

$$\left| D_i L_{k-i}^{q^i} \right| = q^{-\left(\frac{q^i-1}{q-1} + (k-i)q^i\right)}, \quad 0 \leq i \leq k.$$

For large values of k, an elementary investigation of the function $z \mapsto (k-z)q^z$, $z \leq k$, shows that

$$\max_{0 \leq i \leq k} (k-i) q^i \leq \alpha q^k, \quad \alpha > 0,$$

so that

$$|f_k(s)| \leq q^{\alpha q^k}$$

for all $s \in \overline{K}_c$ with $|s| \leq q^{-1}$. Therefore the series (5.38) converges for small $|t|$, so that the Carlitz module function belongs to $\mathcal{F}_2$.

Since $d_s f_i = f_{i-1}$ for $i \geq 1$, and $d_s f_0 = 0$, we see that $d_s C_s(t) = C_s(t)$. Clearly, the function $C_s(t)$ is nonsparse. Therefore the Carlitz module function is quasiholonomic, jointly in both its variables.

(b) *Thakur's hypergeometric polynomials.* We consider the polynomial case of Thakur's hypergeometric function (Section 4.1.2):

$$_lF_\lambda^{(1)}(-a_1,\ldots,-a_l;-b_1,\ldots,-b_\lambda;z) = \sum_m \frac{(-a_1)_m\ldots(-a_l)_m}{(-b_1)_m\ldots(-b_\lambda)_m D_m} z^{q^m} \tag{5.39}$$

where $a_1,\ldots,a_l,b_1,\ldots,b_\lambda \in \mathbb{Z}_+$,

$$(-a)_m = \begin{cases} (-1)^{a-m} L_{a-m}^{-q^m}, & \text{if } m \leq a, \\ 0, & \text{if } m > a, \end{cases} \quad a \in \mathbb{Z}_+. \tag{5.40}$$

It is seen from (5.40) that the terms in (5.39), which make sense and do not vanish, are those with $m \leq \min(a_1,\ldots,a_l,b_1,\ldots,b_\lambda)$. Let

$$f(s,t_1,\ldots,t_l,u_1,\ldots,u_\lambda)$$
$$= \sum_{k_1=0}^\infty \ldots \sum_{k_l=0}^\infty \sum_{\nu_1=0}^\infty \ldots \sum_{\nu_\lambda=0}^\infty {}_lF_\lambda(-k_1,\ldots,-k_l;-\nu_1,\ldots,-\nu_\lambda;s) t_1^{q^{k_1}} \ldots$$
$$\times t_l^{q^{k_l}} u_1^{q^{\nu_1}} \ldots u_\lambda^{q^{\nu_\lambda}}. \tag{5.41}$$

We prove as above that all the series in (5.41) converge near the origin. Thus, $f \in \mathcal{F}_{l+\lambda+1}$.

It is not difficult to check ([111], Section 6.5) that

$$d_s\left({}_lF_\lambda(-k_1,\ldots,-k_l;-\nu_1,\ldots,-\nu_\lambda;s)\right)$$
$$= {}_lF_\lambda(-k_1+1,\ldots,-k_l+1;-\nu_1+1,\ldots,-\nu_\lambda+1;s) \tag{5.42}$$

if all the parameters $k_1,\ldots,k_l,\nu_1,\ldots,\nu_\lambda$ are different from zero. If at least one of them is equal to zero, then the left-hand side of (5.42) equals zero. This property implies the identity $d_s f = f$, the same as that for the Carlitz module function. Since f is nonsparse, it is quasiholonomic.

Next we will see that the K-binomial coefficients (4.52) correspond to a quasiholonomic function satisfying a more complicated equation containing also the operator Δ_t.

(c) *K-binomial coefficients.* By Proposition 4.10, the function

$$f(s,t) = \sum_{k=0}^\infty \sum_{m=0}^k \binom{k}{m}_K s^{q^m} t^{q^k},$$

belonging to $\mathcal{F}_2$, satisfies the equation

$$d_s f(s,t) = \Delta_t f(s,t) + [1]^{1/q} f(s,t).$$

Obviously, f is nonsparse. Therefore f is quasiholonomic.

5.5 Comments

The general framework of filtered and graded rings and modules (Sections 5.1.1, 5.1.2) is well known. See, for example, [16, 30, 79]. Generalized Weyl algebras were introduced by Bavula [10] and studied in his subsequent works, especially in the paper [12] by Bavula and van Oystaeyen.

Investigation of the algebraic structure of the rings of differential operators with the Carlitz derivatives was initiated by the author [63] and continued on a larger scale by Bavula [11] who applied successfully his GWA techniques. Our Section 5.3 is an introductory exposition of part of the results from [11]. Note that Proposition 5.15 (iii) (by Bavula [11]) corrects an erroneous statement from [63].

Properties of differential operators (in the usual sense) over fields of positive characteruistic are quite different; see, for example [87].

Theorem 5.14 is taken from the author's paper [68], which contained also the notion of a quasiholonomic module and the whole related material (in the surveys [69, 70] quasiholonomic modules are called holonomic). The general technique of filtered modules over the Carlitz rings is quite similar to the case of modules over the Weyl algebras, though some of the results are different, in particular due to the existence of nontrivial finite-dimensional modules.

Bibliography

[1] D. Adam, Car-Pólya and Gel'fond theorems for $\mathbb{F}_q[T]$, *Acta Arith.* **115** (2004), 287–303.
[2] S. Albeverio and A. Yu. Khrennikov, Representations of the Weyl group in spaces of square integrable functions with respect to p-adic valued Gaussian distributions, *J. Phys. A: Math. and Gen.* **29** (1996), 5515–5527.
[3] Y. Amice, Interpolation p-adique, *Bull. Soc. Math. France* **92** (1964), 117–180.
[4] Y. Amice, Duals. In: *Proc. Conf. p-Adic Analysis*, Nijmegen, 1978, pp. 1–15.
[5] Y. André and L. Di Visio, q-Difference equations and p-adic local monodromy, *Astérisque* **296** (2004), 55–111.
[6] M. F. Atiyah and I. G. Macdonald, *Introduction to Commutative Algebra*, Addison-Wesley, Reading, 1969.
[7] A. Baker, p-Adic continuous functions on rings of integers and a theorem of K. Mahler, *J. London Math. Soc.* **33** (1986), 414–420.
[8] F. Baldassarri, Differential modules and singular points of p-adic differential equations, *Adv. Math.* **44** (1982), 155–179.
[9] H. Bateman and A. Erdélyi, *Higher Transcendental Functions*, Vol. 2, McGraw-Hill, New York, 1953.
[10] V. Bavula, Generalized Weyl algebras and their representations, *St. Petersburg Math. J.* **4**, No. 1 (1993), 71–92.
[11] V. Bavula, The Carlitz algebras, *J. Pure Appl. Algebra* **212** (2008), 1175–1186.
[12] V. Bavula and F. van Oystaeyen, The simple modules of certain generalized crossed products, *J. Algebra* **194** (1997), 521–566.
[13] M. Bhargava, P-orderings and polynomial functions on arbitrary subsets of Dedekind rings, *J. Reine Angew. Math.* **490** (1997), 101–127.
[14] M. Bhargava, The factorial function and generalizations, *Amer. Math. Monthly* **107** (2000), 783–799.
[15] M. Bhargava and K. S. Kedlaya, Continuous functions on compact subsets of local fields, *Acta Arith.* **91** (1999), 191–198.
[16] J.-E. Björk, *Rings of Differential Operators*, North-Holland, Amsterdam, 1979.
[17] N. Bourbaki, *Algèbre Commutative*, Hermann, Paris, 1961–65.
[18] N. Bourbaki, *Algebra I*, Springer, Berlin, 1989.
[19] A. Buium, Continuous π-adic functions and π-derivations, *J. Number Theory*

84 (2000), 34–39.

[20] P.-J. Cahen and J.-L. Chabert, *Integer-Valued Polynomials*, American Mathematical Society, Providence, 1997.

[21] M. Car, Polya's theorem for $\mathbb{F}_q[T]$, *J. Number Theory* **66** (1997), 148–171.

[22] L. Carlitz, On certain functions connected with polynomials in a Galois field, *Duke Math. J.* **1** (1935), 137–168.

[23] L. Carlitz, A set of polynomials, *Duke Math. J.* **6** (1940), 486–504.

[24] L. Carlitz, Some special functions over $GF(q, x)$, *Duke Math. J.* **27** (1960), 139–158.

[25] P. Cartier, Démonstration "automatique" d'identités et fonctions hypergéometriques (d'après D. Zeilberger), *Astérisque* **206** (1992), 41–91.

[26] D. N. Clark, A note on the p-adic convergence of solutions of linear differential equations, *Proc. Amer. Math. Soc.* **17** (1966), 262–269.

[27] E. A. Coddington and N. Levinson, *Theory of Ordinary Differential Equations*, McGraw-Hill, New York, 1955.

[28] R. F. Coleman, Dilogarithms, regulators and p-adic L-functions, *Invent. Math.* **69** (1982), 171–208.

[29] K. Conrad, The digit principle, *J. Number Theory* **84** (2000), 230–257.

[30] S. C. Coutinho, *A Primer of Algebraic D-modules*, Cambridge University Press, 1995.

[31] L. Delamette, Théorème de Pólya en caractéristique finie, *Acta Arith.* **106** (2003), 159–170.

[32] B. Diarra, The Hopf algebra structure of the space of continuous functions on power series over $\mathbb{F}_q$ and Carlitz polynomials, *Contemp. Math.* **319** (2003), 75–97.

[33] V. G. Drinfeld, Elliptic modules, *Math. USSR Sbornik* **23** (1976), 561–592.

[34] B. Dwork, *Lectures on p-Adic Differential Equations*, Springer, New York, 1982.

[35] B. Dwork, G. Gerotto, and F. J. Sullivan, *An Introduction to G-Functions*, Princeton University Press, 1994.

[36] A. Escassut, *Analytic Elements in p-Adic Analysis*, World Scientific, Singapore, 1995.

[37] L. Ferrari, An umbral calculus over infinite coefficient fields of positive characteristic, *Comp. Math. Appl.* **41** (2001), 1099–1108.

[38] I. B. Fesenko and S. V. Vostokov, *Local Fields and Their Extensions*, American Mathematical Society, Providence, 2002.

[39] F. R. Gantmacher, *Matrizentheorie*, Springer, Berlin, 1986.

[40] E.-U. Gekeler, Some new identities for Bernoulli–Carlitz numbers, *J. Number Theory* **33** (1989), 209–219.

[41] I. M. Gelfand, *Lectures on Linear Algebra*, Interscience, New York, 1961.

[42] A. O. Gelfond, *Calculus of Finite Differences*, Hindustan Publishing Corporation, Dehli, 1971.

[43] D. Goss, Fourier series, measures, and divided power series in the theory of function fields, *K-Theory* **1** (1989), 533–555.

[44] D. Goss, A formal Mellin transform in the arithmetic of function fields, *Trans. Amer. Math. Soc.* **327** (1991), 567–582.

[45] D. Goss, *Basic Structures of Function Field Arithmetic*, Springer, Berlin, 1996.

[46] A. Granville, Arithmetic properties of binomial coefficients. I. In: *Organic*

Mathematics. CMS Conf. Proc. Vol. 20. American Mathematical Society, Providence, 1997, pp. 253–276.
[47] P. Hartman, *Ordinary Differential Equations*, Wiley, New York, 1964.
[48] H. Hasse, Theorie der höheren Differentiale in einem algebraischen Funktionkörper mit vollkommenem Konstantenkörper bei beliebiger Charakterictik, *J. Reine Angew. Math.* **175** (1936), 50–54.
[49] H. Hasse and F. K. Schmidt, Noch eine Begründung der Theorie der höheren Differentialquotienten in einem algebraischen Funktionkörper einer Unbestimmten, *J. Reine Angew. Math.* **177** (1937), 215–237.
[50] I. N. Herstein, *Noncommutative Rings*, Carus Math. Monograph No. 15, Math. Assoc. of America, J. Wiley and Sons, 1968.
[51] E. Hille, *Lectures on Ordinary Differential Equations*, Addison-Wesley, Reading, 1969.
[52] K. Ireland and M. Rosen, *A Classical Introduction to Modern Number Theory*, Springer, New York, 1982.
[53] S. Jeong, A comparison of the Carlitz and digit derivatives bases in function field arithmetic, *J. Number Theory* **84** (2000), 258–275.
[54] S. Jeong, Continuous linear endomorphisms and difference equations over the completion of $\mathbb{F}_q[T]$, *J. Number Theory* **84** (2000), 276–291.
[55] S. Jeong, Hyperdifferential operators and continuous functions on function fields, *J. Number Theory* **89** (2001), 165–178.
[56] H. Keller, H. Ochsenius and W. H. Schikhof, On the commutation relation $AB - BA = I$ for operators on non-classical Hilbert spaces. In *p-Adic Functional Analysis* (A. K. Katsaras et al., eds.), Lect. Notes Pure Appl. Math., Vol. 222, Marcel Dekker, New York, 2001, pp. 177–190.
[57] A. Khrennikov, *Non-Archimedean Analysis: Quantum Paradoxes, Dynamical Systems and Biological Models*, Kluwer, Dordrecht, 1997.
[58] H. Koch, *Galois Theory of p-Extensions*, Springer, Berlin, 2002.
[59] A. N. Kochubei, *Pseudo-Differential Equations and Stochastics over Non-Archimedean Fields*, Marcel Dekker, New York, 2001.
[60] A. N. Kochubei, p-Adic commutation relations, *J. Phys. A: Math. and Gen.* **29** (1996), 6375–6378.
[61] A. N. Kochubei, Harmonic oscillator in characteristic p, *Lett. Math. Phys.* **45** (1998), 11–20.
[62] A. N. Kochubei, $\mathbb{F}_q$-linear calculus over function fields, *J. Number Theory* **76** (1999), 281–300.
[63] A. N. Kochubei, Differential equations for $\mathbb{F}_q$-linear functions, *J. Number Theory* **83** (2000), 137–154.
[64] A. N. Kochubei, Differential equations for $\mathbb{F}_q$-linear functions II: Regular singularity, *Finite Fields Appl.* **9** (2003), 250–266.
[65] A. N. Kochubei, Umbral calculus in positive characteristic, *Adv. Appl. Math.* **34** (2005), 175–191.
[66] A. N. Kochubei, Strongly nonlinear differential equations with Carlitz derivatives over a function field, *Ukrainian Math. J.* **57** (2005), 794–805.
[67] A. N. Kochubei, Polylogarithms and a zeta function for finite places of a function field, *Contemporary Math.* **384** (2005), 157–167.
[68] A. N. Kochubei, Quasi-holonomic modules in positive characteristic, *J. Algebra* **302** (2006), 826–844.
[69] A. N. Kochubei, Umbral calculus and holonomic modules in positive

characteristic, In: *p-Adic Mathematical Physics*, AIP Conference Proceedings, Vol. 826 (2006), 254–266.
[70] A. N. Kochubei, Hypergeometric functions and Carlitz differential equations over function fields, In: *Arithmetic and Geometry around Hypergeometric Functions*, Progress in Mathematics, Vol. 260, Birkhäuser, Basel, 2007, pp. 163–187.
[71] A. N. Kochubei, Evolution equations and functions of hypergeometric type over fields of positive characteristic, *Bull. Belg. Math. Soc.* **14** (2007), 947-959.
[72] A. N. Kochubei, Dwork-Carlitz exponential and overconvergence for additive functions in positive characteristic, *Intern. J. Number Theory* **4** (2008), 453–460.
[73] G. R. Krause and T. H. Lenagan, *Growth of Algebras and Gelfand–Kirillov Dimension*, AMS, Providence, 2000.
[74] S. K. Lando, *Lectures on Generating Functions*, AMS, Providence, 2003.
[75] S. Lang, *Algebra*, Addison-Wesley, Reading, 1965.
[76] R. Lidl and H. Niederreiter, *Finite Fields*, Cambridge University Press, 1983.
[77] E. Lucas. *Théorie des Nombres*, Jacques Gabay, Paris, 1995.
[78] E. Lutz, Sur l'equation $y^2 = x^3 - Ax - B$ dans les corps p-adiques, *J. Reine Angew. Math.* **177** (1937), 238–247.
[79] J. C. McConnell and J. C. Robson, *Noncommutative Noetherian Rings*, AMS, Providence, 2001.
[80] M. B. Nathanson, Additive number theory and the ring of quantum integers In: *General Theory of Information Transfer and Combinatorics*, Lect. Notes Computer Sci. Vol. 4123 (2006), 505–511.
[81] A. Perelomov, *Generalized Coherent States and Their Applications*, Springer, Berlin, 1986.
[82] C. Perez-Garcia, Locally convex spaces over non-Archimedean valued fields, *Contemporary Math.* **319** (2003), 251–279.
[83] R. S. Pierce, *Associative Algebras*, Springer, New York, 1982.
[84] G. Pólya, Über ganzwertige ganze Funktionen, *Rend. Circ. Mat. Palermo* **40** (1915), 1–16.
[85] B. Poonen, Fractional power series and pairings on Drinfeld modules, *J. Amer. Math. Soc.* **9** (1996), 783–812.
[86] M. van der Put, Meromorphic differential equations over valued fields, *Indag. Math.* **42** (1980), 327–332.
[87] M. van der Put, Differential equations in characteristic p, *Compositio Math.* **97** (1995), 227–251.
[88] J. Riordan, *Combinatorial Identities*, Wiley, New York, 1968.
[89] P. Robba and G. Christol, *Équations Différentielles p-Adiques*, Hermann, Paris, 1994.
[90] A. M. Robert, *A Course in p-Adic Analysis*, Springer, New York, 2000.
[91] S. Roman, *The Umbral Calculus*, Academic Press, Orlando, 1984.
[92] S. M. Roman and G.-C. Rota, The umbral calculus, *Adv. Math.* **27** (1978), 95–188.
[93] A. van Rooij, *Non-Archimedean Functional Analysis*, Marcel Dekker, New York, 1978.
[94] M. Rosen, *Number Theory in Function Fields*, Springer, New York, 2002.
[95] G.-C. Rota, D. Kahaner and A. Odlyzko, On the foundations of combinatorial

theory. VIII. Finite operator calculus, *J. Math. Anal. Appl.* **42** (1973), 684–760.
[96] G.-C. Rota and B. D. Taylor, The classical umbral calculus, *SIAM J. Math. Anal.* **25** (1994), 694–711.
[97] S. G. Samko, A. A. Kilbas, and O. I. Marichev, *Fractional Integrals and Derivatives: Theory and Applications*, Gordon and Breach, New York, 1993.
[98] W. Schikhof, *Ultrametric Calculus*, Cambridge University Press, 1984.
[99] F. K. Schmidt, Die Wronskische Determinante in beliebigen differenzierbaren Funktionkörpern, *Math. Z.* **45** (1939), 62–74.
[100] P. Schneider, *Nonarchimedean Functional Analysis*, Springer, Berlin, 2002.
[101] J.-P. Serre, Endomorphismes complètement continus des espaces de Banach p-adiques, *Publ. Math. IHES* **12** (1962), 69–85.
[102] M. Setoyanagi, Note on Clark's theorem for p-adic convergence, *Proc. Amer. Math. Soc.* **125** (1997), 717–721.
[103] D. Sinnou and D. Laurent, Indépendence algebrique sur les T-modules, *Compositio Math.* **122** (2000), 1–22.
[104] S. De Smedt, Some new bases for p-adic continuous functions, *Indag. Math. New Ser.* **4** (1993), 91–98.
[105] B. Snyder. *Hyperdifferential Operators on Function Fields and Their Applications*. PhD thesis, Ohio State University, Columbus, 1999.
[106] A. Sudbery, *Quantum Mechanics and the Particles of Nature*, Cambridge University Press, 1986.
[107] K. Tateyama, Continuous functions on discrete valuation rings, *J. Number Theory* **75** (1999), 23–33.
[108] O. Teichmüller, Differentialrechnung bei Charakteristik p, *J. Reine Angew. Math.* **175** (1936), 89–99.
[109] D. S. Thakur, Hypergeometric functions for function fields, *Finite Fields Appl.* **1** (1995), 219–231.
[110] D. S. Thakur, Hypergeometric functions for function fields II, *J. Ramanujan Math. Soc.* **15** (2000), 43–52.
[111] D. S. Thakur, *Function Field Arithmetic*, World Scientific, Singapore, 2004.
[112] Y. Uchino and T. Satoh, Function field modular forms and higher derivations, *Math. Ann.* **311** (1998), 439–466.
[113] L. Van Hamme, Continuous operators, which commute with translations, on the space of continuous functions on $\mathbb{Z}_p$. In: *p-Adic Functional Analysis* (J. M. Bayod et al., eds.), Lect. Notes Pure Appl. Math. Vol. 137, Marcel Dekker, New York, 1992, pp. 75–88.
[114] A. Verdoodt, Umbral calculus in non-Archimedean analysis. In: *p-Adic Functional Analysis* (A. K. Katsaras et al., eds.), Lect. Notes Pure Appl. Math. Vol. 222, Marcel Dekker, New York, 2001, pp. 309–322.
[115] V. S. Vladimirov, I. V. Volovich and E. I. Zelenov, *p-Adic Analysis and Mathematical Physics*, World Scientific, Singapore, 1994.
[116] A. Volkenborn, Ein p-adisches Integral und seine Anwendungen I, II, *Manuscripta Math.* **7** (1972), 341–373; **12** (1974), 17–46.
[117] J. F. Voloch, Differential operators and interpolation series in power series fields, *J. Number Theory* **71** (1998), 106–108.
[118] C. G. Wagner, Interpolation series for continuous functions on π-adic completions of $GF(q,x)$, *Acta Arithm.* **17** (1971), 389–406.
[119] C. G. Wagner, Linear operators in local fields of prime characteristic, *J.*

Reine Angew. Math. **251** (1971), 153–160.
[120] C. G. Wagner, Interpolation series in local fields of prime characteristic, Duke Math. J. **39** (1972), 203–210.
[121] C. G. Wagner, Differentiability in local fields of prime characteristic, Duke Math. J. **41** (1974), 285–290.
[122] A. Weil, *Basic Number Theory*, Springer, Berlin, 1967.
[123] Z. Yang, *Non-Archimedean Analysis over Function Fields with Positive Characteristic*, PhD thesis, Ohio State University, Columbus, 1999.
[124] Z. Yang, Locally analytic functions over completions of $\mathbf{F}_r(U)$, J. Number Theory **73** (1998), 451–458.
[125] Z. Yang, C^n-functions over completions of $\mathbb{F}_r[T]$ at finite places of $\mathbb{F}_r(T)$, J. Number Theory **108** (2004), 346–374.
[126] J. Yu, Transcendence theory over function fields, Duke Math. J. **52** (1985), 517–527.

Index

C^n-function, 83
K-binomial coefficients, 154
$\mathbb{F}_q$-linear function, 4
π-ordering, 94

Additive Carlitz polynomials, 13
Admissible parameters, 138
Almost normalizing extension, 168

Bargmann–Fock representation, 41, 42, 43
Basic sequence, 62
Bessel–Carlitz functions, 143

Carlitz derivative, 39
Carlitz factorials, 2
Carlitz module function, 39
Coherent states, 41, 42
Completion, 32
Contiguous relations, 141
Cyclic orbit, 171

Degenerate module, 197
Degenerate orbit, 172
Delta operator, 62
Difference quotient, 83
Digit principle, 23
Dimension of a filtered module, 187
Dwork exponential, 160
Dwork–Carlitz exponential, 160

Entire function, 91
Euler function, 179
Extended hyperdifferentiations, 31

Filtered module, 168
Filtered ring, 167
Filtration in a module, 168
Filtration in a ring, 167

Finite place, 33
Formal divided power series, 101
Fractional derivative, 58

General Carlitz polynomials, 24
Generalized binomial polynomial, 94
Generalized factorial, 94
Generalized Taylor formula, 64
Generalized Weyl algebra, 171
Global fields, 32
Good filtration, 170
Graded module, 169
Graded ring, 167
Grading, 167
Grading of a module, 169

Hilbert polynomial, 187
Hyperdifferentiations, 17
hypergeometric equation, 139
hypergeometric function, 138

Induced filtrations, 170
Infinite place, 32
Interpolation polynomials, 75
Invariant operator, 61

Linear orbit, 171
Locally analytic function, 76
Locally constant function, 99

Möbius function, 179
Multiplicity of a filtered module, 187

Non-Archimedean Banach space, 9
Non-Archimedean property, 2
Nonsparse function, 197
Normalized basic sequence, 68
Normalized Carlitz polynomials, 14

Numerical polynomial, 185

Order of an entire function, 91
Orthonormal basis, 9

Place, 32
Pochhammer-type symbol, 138
Prime ideal, 9

Quasiholonomic function, 196
Quasiholonomic module, 193

Reduction, 9
Regular compact set, 75
Residual space, 9
Residue field, 9

Schrödinger representation, 40, 41
Sequence of K-binomial type, 63
Standard filtration, 169
Strongly singular function, 16

Thakur's hypergeometric function, 140
Topological tensor product, 10
Type of an entire function, 91

Ultra-metric inequality, 2

Very well distributed sequence, 76
Volkenborn-type integral, 55

Weight module, 171
Well distributed sequence, 76